AF352919

Advances in Micro and Nano Manufacturing: Process Modeling and Applications

Advances in Micro and Nano Manufacturing: Process Modeling and Applications

Editors

Davide Masato
Giovanni Lucchetta

MDPI • Basel • Beijing • Wuhan • Barcelona • Belgrade • Manchester • Tokyo • Cluj • Tianjin

Editors
Davide Masato
Plastics Engineering
University of Massachusetts
Lowell
Lowell
United States

Giovanni Lucchetta
Industrial Engineering
University of Padova
Padova
Italy

Editorial Office
MDPI
St. Alban-Anlage 66
4052 Basel, Switzerland

This is a reprint of articles from the Special Issue published online in the open access journal *Micromachines* (ISSN 2072-666X) (available at: www.mdpi.com/journal/micromachines/special_issues/Micro_Manufacturing).

For citation purposes, cite each article independently as indicated on the article page online and as indicated below:

LastName, A.A.; LastName, B.B.; LastName, C.C. Article Title. *Journal Name* **Year**, *Volume Number*, Page Range.

ISBN 978-3-0365-3271-4 (Hbk)
ISBN 978-3-0365-3270-7 (PDF)

Contents

About the Editors

Davide Masato

Davide Masato achieved a BS in Mechanical Engineering (2011), an MS in Mechanical Engineering (2014), and a Ph.D. (2018) from the University of Padova in Italy. In 2016, he was a visiting researcher to the University of Bradford (UK), where he worked at the Centre for Polymer Micro and Nano Technology (Polymer MNT). Before joining the faculty at UMass Lowell, he was a post-doctoral researcher at the University of Padova, in the Te.Si. Laboratory for Precision Engineering and Manufacturing. Since 2018, Dr. Masato has been an Assistant Professor in the Department of Plastics Engineering at UMass Lowell, working on plastic product design, sustainable plastics manufacturing, and composites for high-impact resistance. Dr. Masato teaches engineering design, injection mold design, and computer-aided engineering. His research has resulted in over 30 journal publications, over 40 conference presentations, 1 book chapter, 3 patents, and 1 start-up company.

Giovanni Lucchetta

Giovanni Lucchetta is an Associate Professor of Manufacturing Engineering at the University of Padova. His research is focused on the forming technologies of polymeric materials and on micro technologies, with particular reference to precision injection molding. His main research topics are: (1) modeling and characterizing the tribological effects of mold surface properties in the injection micro-molding process; (2) the replication of micro- and nanostructured surfaces; (3) the injection molding of polymer materials reinforced with glass fibers. Combining experimental polymer processing, material characterization and numerical simulations, he has developed and modeled innovative manufacturing technologies for plastics. He has published more than 100 papers in international journals and conference proceedings. The majority of these are related to injection molding and have been published in high-ranking peer-reviewed journals. His academic work is motivated by industry experiences.

Preface to "Advances in Micro and Nano Manufacturing: Process Modeling and Applications"

Micro- and nanomanufacturing technologies have been developed in research and industrial environments to support product miniaturization and the integration of new functionalities. The technological developments of new materials and processing methods have been supported by developing predictive models, which can simulate the interactions between materials, process states, and product properties. Compared with the conventional manufacturing scale, micro- and nanoscale technologies require studying different mechanical, thermal, and fluid dynamics phenomena which need to be assessed and modeled. To fully realize the micro- and nano-scale potential, several challenges still need to be overcome. Among them are understanding material processing behaviors at a reduced scale, developing design criteria, validating process control and monitoring strategies, and defining quality control procedures.

This Special Issue showcases 13 state-of-the-art examples of modeling and simulation in micro- and nanomanufacturing processes. The goal is to provide contemporary examples of the use of modeling and simulation in micro- and nanomanufacturing processes, promoting the diffusion and development of these technologies. The presentation and analysis of new technologies are anticipated to support engineering researchers and practitioners with the development of new micro- and nanomanufacturing products and applications. This book provides the reader with the knowledge to understand different micro- and nanomanufacturing technologies and the associated modeling techniques. The technologies covered in the book include different materials, different processing technologies, and different modeling approaches.

The Guest Editors would like to take this opportunity to thank all the authors for submitting their papers to this Special Issue and for contributing to the diffusion and development of micro- and nanomanufacturing technologies. We would also like to thank all the reviewers for dedicating their time and helping to improve the quality of the submitted papers.

Davide Masato, Giovanni Lucchetta
Editors

micromachines

Editorial for the Special Issue on Advances in Micro and Nano Manufacturing: Process Modeling and Applications

Davide Masato [1],* and **Giovanni Lucchetta** [2],*

1. Department of Plastics Engineering, University of Massachusetts Lowell, 1 University Avenue, Lowell, MA 01854, USA
2. Department of Industrial Engineering, University of Padova, Via Venezia 1, 35131 Padova, Italy
* Correspondence: Davide_Masato@uml.edu (D.M.); giovanni.lucchetta@unipd.it (G.L.)

Citation: Masato, D.; Lucchetta, G. Editorial for the Special Issue on Advances in Micro and Nano Manufacturing: Process Modeling and Applications. *Micromachines* **2021**, *12*, 970. https://doi.org/10.3390/mi12080970

Received: 11 August 2021
Accepted: 12 August 2021
Published: 16 August 2021

Publisher's Note: MDPI stays neutral with regard to jurisdictional claims in published maps and institutional affiliations.

Micro- and nano-manufacturing technologies have been developed in research and industrial environments to support product miniaturization and the integration of new functionalities. The technological development of new materials and processing methods has been supported by developing predictive models, which can simulate the interactions between materials, process states, and product properties. Compared with the conventional manufacturing scale, micro- and nano-scale technologies require different mechanical, thermal, and fluid dynamics phenomena to be studied and modeled. To fully realize the micro- and nano-scale potential, several challenges still need to be overcome. Among them are understanding the material processing behavior at a reduced scale, developing design criteria, validating process control and monitoring strategies, and defining quality control procedures.

This Special Issue showcases 12 state-of-the-art examples of modeling and simulation in micro- and nano-manufacturing processes. The papers that are published in this Special Issue explore the following micro- and nano-manufacturing processes: laser texturing [1,2], 3D printing [3,4], grinding [5,6], turning [7,8], and electrochemical machining [9,10]. The remaining papers cover lithography-based manufacturing [11], and micro- and nano-scale design aspects [12].

Laser texturing technologies exploit electromagnetic energy, to melt and vaporize the material on the surface, creating a texture. Ultrafast texturing is a recent development in laser technologies, which has many peculiarities and allows the machining of different textures. Ultrafast lasers are pulsed lasers that are characterized by a duration of a few nanoseconds, and a frequency ranging between one and several hundreds of kHz. Ultrafast laser systems are characterized by low-power characteristics (i.e., units of Watts); however, the use of high-power light pulses allows for high values of energy to be focused on the workpiece surface. Kažukauskas et al. used a femtosecond laser to polish the surface of a transparent biocompatible polymer for body implants [1]. The results showed that the polishing process depends on the thermal heat flux, which initiates thermos-dynamical phenomena that change the surface roughness of the sample. Piccolo et al. used a femtosecond laser to create a submicron texture on a metallic mold, which was then replicated by microinjection molding [2]. The texture replication resulted in the functionalization of a plastic surface, which showed different static wetting behavior, as a function of the texture replication attained with the molding process.

Recent innovations in 3D printing technologies allow the direct manufacturing of micro- and nano-scale products, such as microfluidics, and the generation of fine details on manufacturing tools. Dempsey et al. used a digital light processing (DLP)-based inverted stereolithography process, to produce thermoset polymer-based tooling for micro injection molding [3]. The results show that the resistance of the soft tool is related to its dynamic temperature distribution during the process, and its ability to dissipate the heat that is convected by the hot polymer melt. Chen et al. developed a sample library of standardized components and connectors, manufactured using a multi-jet modeling technology, using a print head that jets the photopolymer and the waxy support material [4]. Their modular microfluidic system offers potential for realizing mass-production complexes and multiplex systems for the commercialization of microfluidic platforms.

Grinding, at the nano-scale level, can be efficiently employed to create surfaces with ultrahigh precision, by removing a few atomic layers from the substrate. Manea et al developed an arithmetical model to calculate the nano-scale grinding force required to machine optical glass, and analyzed how processing parameters can be controlled to achieve high surface and subsurface quality [5]. Karkalos et al. focused on molecular dynamics modeling with multiple abrasive grains, to study the effect of spacing between the adjacent rows of abrasive grains and the effect of the rake angle of the abrasive grains [6]. They evaluated the grinding forces and temperatures, ground surface quality, chip formation, and subsurface damage of the substrate.

Skrzyniarz studied the minimum chip thickness formation during the longitudinal turning of steel [7]. The analysis of the machined workpiece showed that the depth of cut and the feed rate affects the minimum chip thickness. Elastic-plastic deformation and ploughing were observed when the feed rate was lower than the cutting edge radius. You et al. carried out an experimental investigation, focusing on laser-assisted turning by the in-process heating of tungsten carbide, for precision glass molding [8]. The process guides the laser beam passes through the transparent tool, to heat the material locally, thereby decreasing the ultra-high tungsten carbide hardness and altering the fracture toughness as soon as the tool interacts with the material. Through an experimental study, the technology was shown to improve machinability, making the ductile turning of hard and brittle materials possible, and has been confirmed as a viable solution for hard and brittle material ultra-precision machining.

Electrochemical machining uses energy (i.e., chemical, electrical, thermal) to remove material from the workpiece's surface, without physical contact. He et al. studied radial ultrasonic rolling electrochemical micromachining, which is a process that is characterized by the feeding of small and rotating electrodes, aided by ultrasonic rolling, to create an array of pits [9]. Fan et al. analyzed a method of electrochemical machining, using a conductive masked porous cathode and jet electrolyte supply, to generate micro-grooves with high machining localization [10].

A lithography process involves transferring a geometric pattern from a mask to a photoresist-coated semiconductor wafer surface. Geng et al. used a high-precision lithography simulation model for thick SU-8 photoresist, based on the waveguide method, to calculate light intensity in the photoresist and predict the profiles of developed SU-8 structures [11].

Langford et al. developed a methodology to calculate the forces acting upon the intraosseous transcutaneous amputation prosthesis, which was designed for use in a quarter amputated femur [12]. The design approach focused on a failure feature, shaped similarly to a safety notch, which would stop excessive stress permeating the bone, causing damage to the user. The topology analysis identified critical materials and local maximum stresses when modeling the applied loads.

The guest editors would like to take this opportunity to thank all the authors for submitting their papers to this Special Issue, and for contributing to the diffusion and development of micro- and nano-manufacturing technologies. We would also like to thank all the reviewers for dedicating their time and helping to improve the quality of the submitted papers.

Conflicts of Interest: The author declares no conflict of interest.

References

1. Kažukauskas, E.; Butkus, S.; Tokarski, P.; Jukna, V.; Barkauskas, M.; Sirutkaitis, V. Micromachining of Transparent Biocompatible Polymers Applied in Medicine Using Bursts of Femtosecond Laser Pulses. *Micromachines* **2020**, *11*, 1093. [CrossRef] [PubMed]
2. Piccolo, L.; Sorgato, M.; Batal, A.; Dimov, S.; Lucchetta, G.; Masato, D. Functionalization of Plastic Parts by Replication of Variable Pitch Laser-Induced Periodic Surface Structures. *Micromachines* **2020**, *11*, 429. [CrossRef] [PubMed]
3. Dempsey, D.; McDonald, S.; Masato, D.; Barry, C. Characterization of Stereolithography Printed Soft Tooling for Micro Injection Molding. *Micromachines* **2020**, *11*, 819. [CrossRef] [PubMed]
4. Chen, X.; Mo, D.; Gong, M. 3D Printed Reconfigurable Modular Microfluidic System for Generating Gel Microspheres. *Micromachines* **2020**, *11*, 224. [CrossRef] [PubMed]
5. Manea, H.; Cheng, X.; Ling, S.; Zheng, G.; Li, Y.; Gao, X. Model for Predicting the Micro-Grinding Force of K9 Glass Based on Material Removal Mechanisms. *Micromachines* **2020**, *11*, 969. [CrossRef] [PubMed]
6. Karkalos, N.E.; Markopoulos, A.P. Molecular Dynamics Study of the Effect of Abrasive Grains Orientation and Spacing during Nanogrinding. *Micromachines* **2020**, *11*, 712. [CrossRef] [PubMed]
7. Skrzyniarz, M. A Method to Determine the Minimum Chip Thickness during Longitudinal Turning. *Micromachines* **2020**, *11*, 1029. [CrossRef] [PubMed]
8. You, K.; Fang, F.; Yan, G.; Zhang, Y. Experimental Investigation on Laser Assisted Diamond Turning of Binderless Tungsten Carbide by In-Process Heating. *Micromachines* **2020**, *11*, 1104. [CrossRef] [PubMed]
9. He, K.; Chen, X.; Wang, M. The Optimal Processing Parameters of Radial Ultrasonic Rolling Electrochemical Micromachining—RSM Approach. *Micromachines* **2020**, *11*, 1002. [CrossRef] [PubMed]
10. Fan, G.; Chen, X.; Saxena, K.K.; Liu, J.; Guo, Z. Jet Electrochemical Micromachining of Micro-Grooves with Conductive-Masked Porous Cathode. *Micromachines* **2020**, *11*, 557. [CrossRef] [PubMed]
11. Geng, Z.-C.; Zhou, Z.-F.; Dai, H.; Huang, Q.-A. A 2D Waveguide Method for Lithography Simulation of Thick SU-8 Photoresist. *Micromachines* **2020**, *11*, 972. [CrossRef] [PubMed]
12. Langford, E.; Griffiths, C.; Rees, A.; Bird, J. The Micro Topology and Statistical Analysis of the Forces of Walking and Failure of an ITAP in a Femur. *Micromachines* **2021**, *12*, 298. [CrossRef] [PubMed]

Article

Micromachining of Transparent Biocompatible Polymers Applied in Medicine Using Bursts of Femtosecond Laser Pulses

Evaldas Kažukauskas [1,*], Simas Butkus [1], Piotr Tokarski [2], Vytautas Jukna [1], Martynas Barkauskas [1,3] and Valdas Sirutkaitis [1] 

[1] Laser Research Center, Faculty of Physics, Vilnius University, Saulėtekio Ave. 10, LT-10223 Vilnius, Lithuania; simas.butkus@ff.vu.lt (S.B.); vytautas.jukna@ff.vu.lt (V.J.); martynas.barkauskas@ff.vu.lt (M.B.); valdas.sirutkaitis@ff.vu.lt (V.S.)
[2] Msl Med Services Ltd., Rodou 6, Tremithousa, Paphos 8270, Cyprus; tokarski.ils.germany@gmail.com
[3] Light Conversion, Keramikų 2b, LT-10233 Vilnius, Lithuania
* Correspondence: evaldas.kazukauskas@ff.vu.lt

Received: 22 November 2020; Accepted: 8 December 2020; Published: 10 December 2020

Abstract: Biocompatible polymers are used for many different purposes (catheters, artificial heart components, dentistry products, etc.). An important field for biocompatible polymers is the production of vision implants known as intraocular lenses or custom-shape contact lenses. Typically, curved surfaces are manufactured by mechanical means such as milling, turning or lathe cutting. The 2.5 D objects/surfaces can also be manufactured by means of laser micromachining; however, due to the nature of light–matter interaction, it is difficult to produce a surface finish with surface roughness values lower than ~1 µm Ra. Therefore, laser micromachining alone can't produce the final parts with optical-grade quality. Laser machined surfaces may be polished via mechanical methods; however, the process may take up to several days, which makes the production of implants economically challenging. The aim of this study is the investigation of the polishing capabilities of rough (~1 µm Ra) hydrophilic acrylic surfaces using bursts of femtosecond laser pulses. By changing different laser parameters, it was possible to find a regime where the surface roughness can be minimized to 18 nm Ra, while the polishing of the entire part takes a matter of seconds. The produced surface demonstrates a transparent appearance and the process shows great promise towards commercial fabrication of low surface roughness custom-shape optics.

Keywords: femtosecond micromachining; burst processing; intraocular lens; hydrophilic acrylic; surface roughness; polishing

1. Introduction

Vision is one of the most important senses, which enables us to experience and understand our surroundings. However, the eye, just as any other organ or tissue in the human body, deteriorates with age and vision deterioration substantially contributes to the decrease in the quality of life. According to statistics, about 50% of Americans have developed cataracts by the age 75–79 [1] and every third person in the world suffers from myopia [2]. Today's advanced technologies provide means to cure the above-mentioned eye conditions by changing the natural lens of the eye with vision implant known as intraocular lenses (IOL). However, the manufacturing of intraocular lenses remains time consuming and an economically challenging procedure to this day, this is especially true when considering customized IOLs.

Intraocular lenses are typically manufactured using lathe cutting, milling, compression molding or injection molding [3]. All of the above-mentioned methods use mechanical tools that have their downsides. Diamond tools used for lathe cutting are expensive, need constant recalibration and wear out with time; therefore, leading to the increase of the surface roughness of the produced part. In addition, milling of the lens contour is a time intensive process taking up to five minutes per lens, the calibration of the equipment can take up to two hours daily [4]. Other methods that include melting of plastic and then molding are more suitable for mass production as they allow production of large quantities of units in a short period of time. Even though molding is great for mass production, this technology is not suited for the production of customized lenses, as every customized lens requires a new custom mold. The manufacturing of molds requires high precision machinery; therefore, the molds are expensive and difficult to manufacture [5]. Despite which method is used for the manufacturing of the lenses, the surfaces typically do not satisfy optical-quality standards and require an additional polishing procedure to be carried out. Both sides of the lens need to be polished in order to reach optical-quality standards (<10 nm Ra). Polishing methods such as pitch polishing, polishing using synthetic pads or magneto rheological figuring [6] are not suitable for intraocular lens production as there is a high risk of damaging the lens haptics. To this day all intraocular lenses are polished using the tumbling method, where lenses are tumbled in a mixture of glass beads, alcohol and deionized water. This method is capable of producing surfaces of superb quality, in other words, Ra in the range of a few nanometers [7]; however, the tumbling process may take as long as few days [8]. To avoid the stated disadvantages of mechanical manufacturing, laser micromachining could be used as an alternative to mechanical manufacturing means of intraocular or other type of optical lenses. Nowadays it is possible to achieve high ablation efficiencies of 1.23 μm/pulse using Ti:Sapphire laser pulses of 30 μJ energy in polymer poly(methyl methacrylate), better known as PMMA [9]. In addition, the surface quality achieved may be in the range of a few hundred nm to ~1 μm depending on wavelength used [10]. This is achieved thanks to the advanced technologies of ultrashort pulse generation in the pico and femto duration range. The peak intensity of focused ultrashort pulses can be as high as 10^{13} W/cm^2 enabling the initiation of non-linear processes such as multi-photon absorption and avalanche ionization that enable the micromachining of materials that under normal circumstances (visible to infrared spectrum) are transparent. When using femtosecond laser pulses the material is instantly vaporized, resulting in relatively "cold" ablation regimes without producing heat-affected zones (also known as HAZ) [11]. The utilization of ultrashort infrared femtosecond pulses in the field of laser micromachining provides the flexibility to produce almost any shape from transparent or non-transparent materials. However, the non-linear absorption of the pulse and accompanied instant material evaporation alone is not sufficient to produce an optical-grade quality surface finish resulting in surface roughness >1 μm Ra. Transparent surfaces with roughness of >1 μm exhibit strong light scattering and may reduce imaging quality of these components to such a degree that it becomes too low for optical applications. Surface roughness is also important for the bio-compatibility of the implants [12–14]. Rough surfaces on the implants can damage human tissues or be rejected by the immune system. Moreover, it was seen that implants with higher surface roughness values increase the risk of post–surgical complications (e.g., posterior capsular opacification) that lead to decrease in vision [15]. Recently it was discovered that using bursts of femtosecond laser pulses, when each pulse is divided into a sequence of sub-pulses with a temporal separation of few tens of nanoseconds or hundreds picoseconds, improves laser material processing by boosting the ablation efficiency [16–21]. It was also demonstrated, that by using bursts of pulses it is possible to further decrease the surface roughness compared to using the conventional single femtosecond pulse regimes [22].

The aim of this study is the investigation of the polishing capabilities of rough (~1 μm Ra) hydrophilic acrylic (typical material used for vision implants) surfaces using bursts of femtosecond laser pulses, while the initial rough surface of the samples was prepared by femtosecond laser ablation to a desired shape without using bursts. It was shown, that by tailoring the properties of the burst (the shape of the envelope, number of pulses, average power, etc.) it is possible to control the amount of heat flux entering the material and in such a way realize a controlled melting procedure that reduces the surface roughness from ~1 μm Ra to <20 nm Ra. These results show great potential for the industrial scale production of customized optical components made from transparent, biocompatible polymer materials.

2. Materials and Methods

In this work we used a multi-burst femtosecond laser "Carbide" (Light Conversion, Vilnius, Lithuania). With a central wavelength of 1030 nm, generating pulses of 220 fs (full width at half maximum) duration at a repetition rate of 100 kHz, with a maximum average power of 40 W. The laser is able to produce controllable bursts of pulses, meaning that each pulse may be divided into a sequence of sub-pulses with a variable energy distributed through the sub-pulses as shown in Figure 1. The temporal separation between the sub-pulses may be set to either 400 ps or 15.5 ns. In the scope of this article, only the temporal separation of 400 ps was investigated. The formation of femtosecond bursts is done by storing pulses in the regenerative amplifier and controllably expelling them. The control of the expelling pulse energy is done by using electro-optical switches. The precise timing of the round trip of the regenerative amplifier and the time delay between the pulses from the master oscillator is crucial [23]. In addition, the laser has a built-in software feature that allows tailoring of the energy distribution of the sub-pulses within the burst which can produce an ascending or descending burst amplitude envelope, meaning it is possible to form a rising/falling energy envelope, see Figure 1. To characterize the energy of the sub-pulses within the burst we used the "Time-correlated Single Photon Counting" (also known as TCSPC) method [24] and MATLAB (R2018a version) for data visualization. The micromachining of the material was done by focusing the beam via a telecentric 100 mm F-theta lens, and controlling the position of the beam on the surface of the sample using a galvanoscanner "intelliSCAN 14" (SCANLAB, Munich, Germany), the setup is shown in Figure 1. In addition, the scanner with the lens was mounted on a z-axis translation stage, thus enabling the manipulation of the beam in x, y and z directions. The position of the beam on the sample was controlled using micromachining software DMC (1.4.71 version). The sample was precisely positioned below the scanner using a firmly fixed custom-built sample holder. The samples were Contamac CONTAFLEX 26% UV-IOL (R) hydrophilic acrylic tablets having diameter of 15 mm and thickness of 3 mm. The beam waist was positioned directly on the surface of the sample and the focal position was fine-tuned by firing single pulses at the sample's surface at different heights with a step of 100 μm until the smallest diameter crater was achieved. The produced craters were investigated using an "Olympus BX51" (OLYMPUS, Hamburg, Germany) microscope. We determined that using an average laser power of 2.5 W (fluence—6.5 J/cm^2) the diameter of the crater at the focal position was 20.7 μm, while using an average laser power of 33.6 W (fluence—88 J/cm^2) the diameter of the crater was 77.3 μm. To determine the material ablation threshold we used a method described in [25] that consists of several steps: first, the square of the ablated crater diameter is plotted versus the logarithm of the pulse energy. From the slope of the distribution the focal spot $\omega_0(1/e^2)$ can be calculated. Having the spot size, the fluence was calculated and crater diameter squared versus the logarithm of fluence was plotted. By extrapolating the dependence to zero the ablation threshold was determined and was found to be 2.7 J/cm^2.

The experiment consisted mainly of two parts: first-characterization of the burst using the TCSPC setup and second-analysis of the micromachined sample. We used the characterization of the burst pulses to find the correct settings of the laser so that each sub-pulse in the burst would have the same energy. To achieve that we placed a scattering surface at an angle of 45° obstructing the beam and collecting the scattered light using a semiconductor detector "ID100-20 visible Single-photon detector" (ID QUANTIQUE, Geneva, Switzerland) and a multichannel electronic system "PicoHarp 300" (PicoQuant, Berlin, Germany). The bursts were registered and visualized using data visualization software. The temporal resolution of the imaging was 50 ps, which enables the accurate representation of the energies of the sub-pulses within the burst. The scattering surface was necessary for decreasing the intensity of the laser beam so as not to damage the detector and avoid possible reflections from the optical surfaces that could influence the registered signals. The parameter that controls the burst energy envelope was changed from −1 to 1 (see different value meanings in Figure 1) for different burst packets consisting of 2, 5, 10 and 25 sub-pulses per burst. By changing the burst envelope parameter, different configurations of sub-pulse amplitudes were obtained and analyzed.

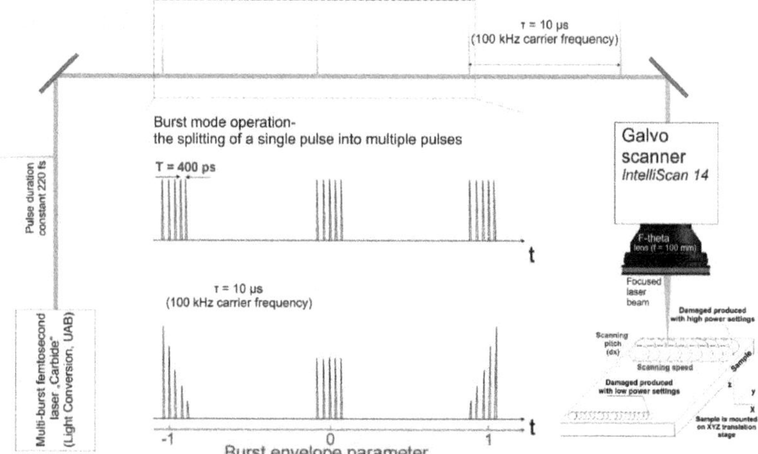

Figure 1. Micromachining setup used for ablation and subsequent polishing of plastic materials. A single pulse may be split into a sequence of sub-pulses (bursts), the temporal separation between each sub-pulse within the burst was constant at 400 ps. Additionally, by varying the burst envelope parameter it is possible to adjust the energy distribution within the burst.

Micromachining of plastic experiments were carried out using a burst envelope parameter value at which all sub-pulses had the same energy. A cuboid having dimensions of 10 mm × 10 mm × 0.5 mm (such dimensions are typical for eye implants) was ablated using the conventional femtosecond ablation regime (single pulses) when the laser was operating at 100 kHz. For ablation of the surface, the overlap of beams was kept constant at 50% in both x and y directions to ensure uniform surface modification over the whole area. After the ablation step a series of experiments were performed, where combinations of bursts with different parameter settings were used for polishing of the ablated surface, see Figure 2B. In the framework of this article, surface polishing is regarded as a sensitive thermodynamical process that is based on remelting of a thin surface layer. While the material is in the liquid state it is pulled in the directions where the surface tensile force is strongest, meaning that it is pulled mainly into various valleys and cavities, hence the surface roughness after solidification may get smaller if the correct laser heating parameters are found. In addition, this process results in a combination of remelting of a thin surface layer smaller than 5 µm and vaporization of micro edges [26–30], see Figure 2A. Using laser radiation, it is possible to enhance the surface quality of many different materials such as metals, plastics, ceramics, glasses, etc. Hence, laser polishing is a highly used technique for industrial applications [31–33].

In our study the surface polishing experiments were conducted using a galvo-scanner based scanning/focusing system. By changing the average laser power (P), the scanning velocity of the beam on the surface of the sample (v), the line scanning pitch (dx) and the number of sub-pulses within the burst (N_p), the polishing capabilities of the plastic material were investigated in order to achieve the best surface quality. While conducting micromachining of samples, the temperature at the surface was monitored using a thermal vision camera "FLIR A600-Series" (FLIR systems, Hoogstraten, Belgium); for image and video processing we used the "FLIR ResearchIR" (ResearchIR MAX 4 version) software package. We also examined how the initial surface roughness impacts the polishing process. Therefore, we conducted a parametric study, where multiple samples were ablated with different settings so as to produce a different surface finish (roughness values ranging from 0.7 to 3 µm) and evaluated its roughness after polishing. In addition, a sample having roughness of Ra = 40 nm, which was prepared by mechanical polishing, was also studied. The "Sensofar PLU 2300" optical profiler (SENSOFAR, Terrassa, Spain) was used for the investigation of the surface topography while the bright field images were taken with the "Olympus BX51" (OLYMPUS, Hamburg, Germany) (10× magnification) and "Olympus LEXT OLS5000" (10× magnification) microscopes (OLYMPUS, Hamburg, Germany). The roughness parameter R_a (arithmetic average height) was chosen as an evaluation criterion of the surface roughness. It is defined as an average absolute deviation of the height from the mean line over one sampling length [34] (in our case sampling length was 125 µm). To process the profiler data, "SensoMap" software was used.

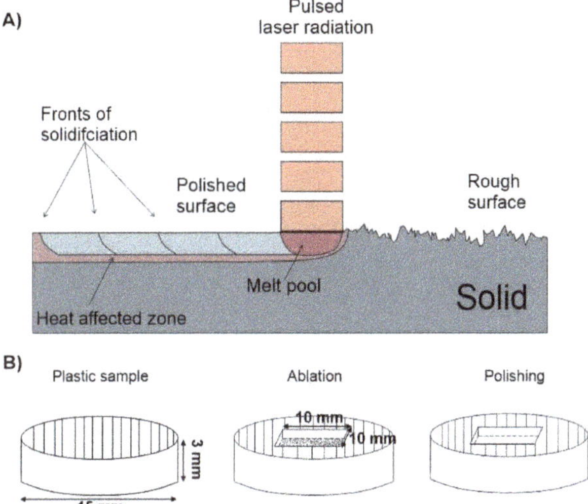

Figure 2. Laser polishing using pulsed laser radiation scheme (**A**), micromachining steps used for polishing (**B**). The structure of 10 mm × 10 mm × 0.5 mm was ablated and then polished using bursts of femtosecond pulses.

3. Results and Discussion

3.1. Burst Investigation Results

Measuring the individual amplitudes of each sub-pulse within the burst is complicated with conventional photodiodes and oscilloscopes due to the insufficient temporal resolution; therefore, the TCSPC method was used. The main reason for these measurements was to adjust the burst envelope parameter in such a way that all of the energies of the sub-pulses within the burst are equal or as close to equal as possible. It was found that tailoring of the burst envelope parameter was necessary to produce a flat envelope for a different number of sub-pulses within the burst. Knowing the individual energy of the sub-pulses is important for experiment reproducibility and for comparison purposes with different burst settings. The experiments were carried out by dividing one pulse into packets of 2, 5, 10 and 25 sub-pulses and different sub-pulse amplitudes were produced when changing the burst envelope parameter. It was determined that splitting a single pulse into a burst is done efficiently up to 10 sub-pulses. Splitting the pulse into more than 10 sub-pulses causes formation of parasitic pre-pulses, see Figure 3B,C. The true origin of the pre-pulses remains unknown, as it is a direct cause of the laser construction. Analysis of the measurement data showed that it is possible to level all sub-pulses within the burst to the same energy with an error of 10% except for the last sub-pulse. However, if the number of sub-pulses within the burst is lower than five, then all of the pulses within the burst are leveled to an 10% error margin. If the number of pulses is greater than five, then the trailing sub-pulse has a higher energy compared to the other sub-pulses (the more sub-pulses in the burst, the higher the energy of the last sub-pulse), this phenomenon is an inherent laser feature. Another interesting insight is that the burst envelope parameter takes on different values for a different number of sub-pulses. It was expected, that the burst envelope parameter value of 0 would yield a burst with equal sub-pulse amplitudes; however, it was found that the envelope appears to be slightly shifted towards an ascending shape (amplitudes of the sub-pulses are increasing). When performing polishing experiments, the burst envelope parameter was varied for the different cases in order to produce a flat envelope.

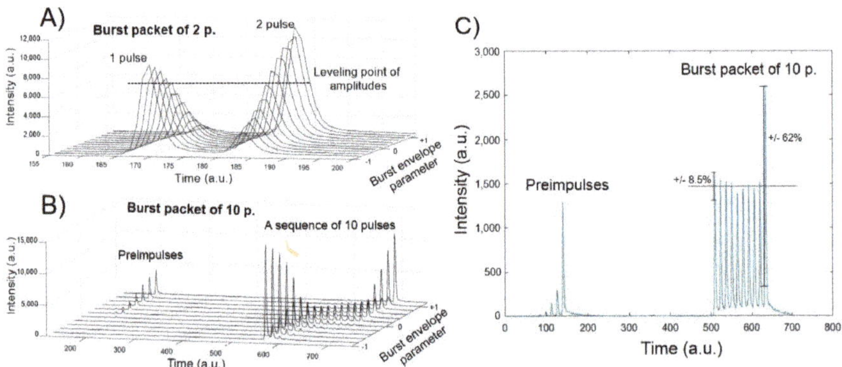

Figure 3. The dependence of the intensity of the sub-pulses in the burst versus the envelope parameter when a single pulse is divided into 2 sub-pulses (**A**) and into 10 sub-pulses (**B**). Amplitudes of a leveled burst packet of 10 sub-pulses are displayed in (**C**).

3.2. Plastic Ablation Results and Polishing Capabilities of Bursts of Femtosecond Laser Pulses

Different materials ablated with femtosecond pulses is not a new topic, therefore an investigation regarding the ablation efficiency when working with different laser parameters was not performed. The preparation of the samples was done with the same laser system in single pulse mode. The ablation threshold was determined for hydrophilic acrylic using the method described in the previous chapter and was found to be 2.7 J/cm^2, see Figure 4B. The determined ablation threshold agrees with the results published by other parties when similar materials were investigated. Nam et.al. [35] and Baudach et.al. [36] reported thresholds of 2.4 J/cm^2 and 2.6 J/cm^2, respectively, while using Ti:Sapphire laser of λ = 800 nm wavelength and pulse duration of τ = 150 fs. On the other hand, Heberle et.al. [10] reported ablation thresholds of Contaflex yellow and Contaflex stan (PMMA materials from the supplier Contaflex) to be 12.9 J/cm^2 and 13.9 J/cm^2, respectively, for a wavelength of λ = 1064 nm and pulse duration of τ = 10 ps. The difference in ablation threshold can be explained by the different pulse duration.

The ablation rate was determined for different pulse energies when the scanning velocity, the scanning pitch and the repetition rate were fixed to 3.6 m/s, 36.57 µm and 100 kHz, respectively. The maximum ablation rate was found to be 0.84 mm^3/s when the fluence is set to 78.5 J/cm^2, which corresponds to an average laser power of 29 W, see Figure 4C. In this case it took approximately 1 min to ablate the 10 mm × 10 mm × 0.5 mm (50 mm^3) shape. It is worth mentioning, that 50 mm^3 is a common value which would be required to produce a custom 3D lens shape applicable for eye implants; therefore, it represents a realistic estimation of the time needed for the ablation task. After ablating the samples to a depth of 0.5 mm, one extra scan was performed with a low average power setting of 2.5 W to further decrease surface roughness. This way we were able to achieve the lowest surface roughness of Ra = 0.991 µm using the single pulse regime.

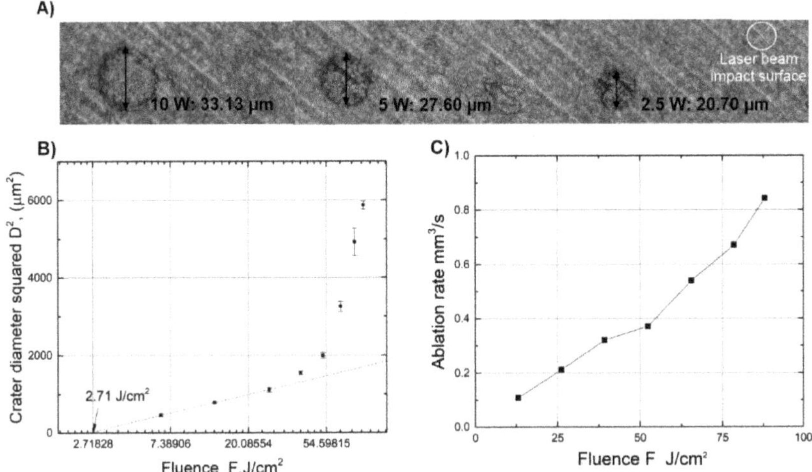

Figure 4. Bright field images of ablated craters taken with "Olympus BX51" microscope (**A**). Crater diameter squared dependence on fluence for ablation threshold retrieval (**B**). Ablation rate dependence on beam fluence (**C**).

The next step in lens production would be to initiate polishing of the produced samples by means of controlled surface melting using bursts of femtosecond pulses while using the same laser system setup. We have used bursts of femtosecond pulses where the sub-pulses had a temporal separation of 400 ps. From previous investigations [21] it was known, that the heat input on the surface would be much more pronounced in this time scale as compared to a larger temporal separation (nanoseconds) setting. However, it was unknown what the best parameter combination that produces the highest heat flux onto the surface is. In our experiments we varied the average laser power, the scanning speed, the scanning pitch and the number of sub-pulses within the burst and analyzed the roughness of the polished surface. The summarized results are shown in Figure 5. The results show that, under certain burst pulse parameter settings, it is possible to reduce the initial roughness of the sample made with the single pulse ablation mode. At 5 sub-pulses in the burst and low average power (~15 W) settings the achieved reduction in surface roughness is already about 2.5-fold, from ~1 μm to ~400 nm. We found that with higher average power settings (~24 W) it was possible to reach a surface roughness value below 100 nm (Ra). This result shows that depending on the scanning speed and spatial pitch settings a significant amount of laser power is required to be pointed onto the surface to initiate melting and surface smoothening. On the other hand, too much power can result in increase of surface roughness when the scanning speed is too high. As evident from the roughness dependence on the scanning speed graph, the surface roughness increases as the scanning speed increases, leading to insufficient thermal accumulation on the surface. However, it was noticed that if the thermal input is too large, boiling of the material can occur and large craters form on the surface producing a rise in the surface roughness as in the case of the low scanning pitch (5 μm) setting, see Figure 5A. It is likely, that the same boiling of the material will appear when the scanning speed would be decreased below the investigated range; however, since one of the goals is to produce rapid and high-quality polishing, low scanning speed settings were not investigated. The results suggest that depending on the used parameter configuration the surface quality after ablation can be enhanced up to >10-fold as can be seen in Figure 6C, or degraded by >3-fold, see Figure 6D. The transparency of the surface is dependent on the scattering properties of the surface. The Rayleigh roughness criterion R_{at} [37] may be used to evaluate the scattering characteristics of the surface when light is incident on the surface:

$$R_{at} = k_0 S_q \frac{|n_1 \cos \theta_i - n_2 \cos \theta_t|}{2},$$

(1)

where k_0—incident wave vector, S_q—surface RMS roughness, $n_{1,2}$—refractive indexes of incident mediums and $\theta_{i,t}$—incident and refracted. Generally, a surface may be regarded as flat with little to no scattering losses if $R_{at} < \pi/16$. Parts A and B appear opaque since R_{at} is higher by approximately an order of magnitude compared to the specified value in this case. Part C in Figure 6 is below the designated criterion and therefore appears transparent. An exception should be applied for part D where the surface appears transparent though shows large roughness values; this is due to the waviness that occurs on the surface when the hot bubbles explode upon reaching the surface. A wavy structure appears on the surface having smooth peaks and valleys. Such surfaces (part D) are not potential candidates for optical applications due to wavefront distortions that arise from the waviness. Overall, it was seen that the parameter space in which the surface roughness decreases drastically is relatively small and the best surface roughness of Ra = 60.8 nm was achieved using the following parameters: P = 25.8 W, v = 500 mm/s, dx = 10 µm and N_p = 10. This demonstrates that using these parameter settings we can achieve optimal thermodynamical flow, hence only melting the thin upper layer of the sample. Additionally, similar results (Ra = 83 nm) were also achieved using the following parameters: P = 25.8 W, v = 1000 mm/s, dx = 10 µm and N_p = 5. This indicates that it is possible to achieve surface smoothening even with a lower than 10 number of sub-pulses in the burst.

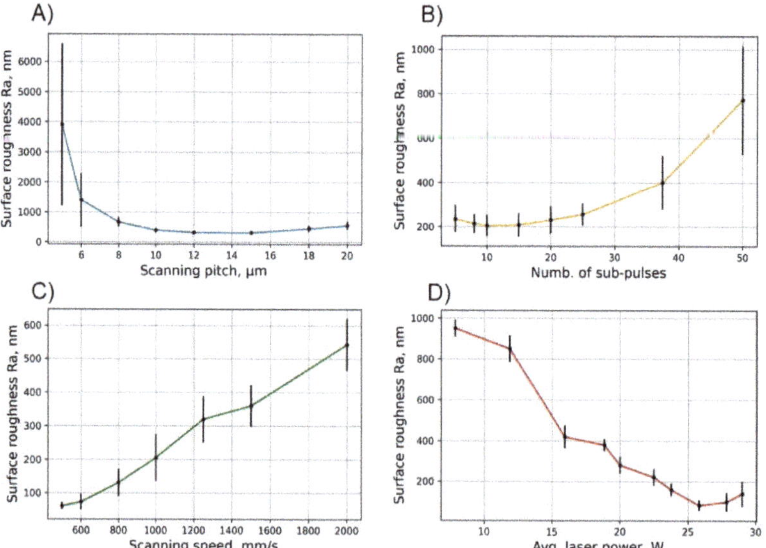

Figure 5. Surface roughness after polishing ablated samples dependent on different polishing parameters. Other fixed parameters in case of (**A**): P = 29 W, N_p = 10 and v = 1000 mm/s; in case of (**B**): P = 19 W, dx = 10 µm and v = 1000 mm/s; in case of (**C**): P = 25.8 W, dx = 10 µm and N_p = 10; and in case of (**D**): v = 1000 mm/s, dx = 10 µm and N_p = 5.

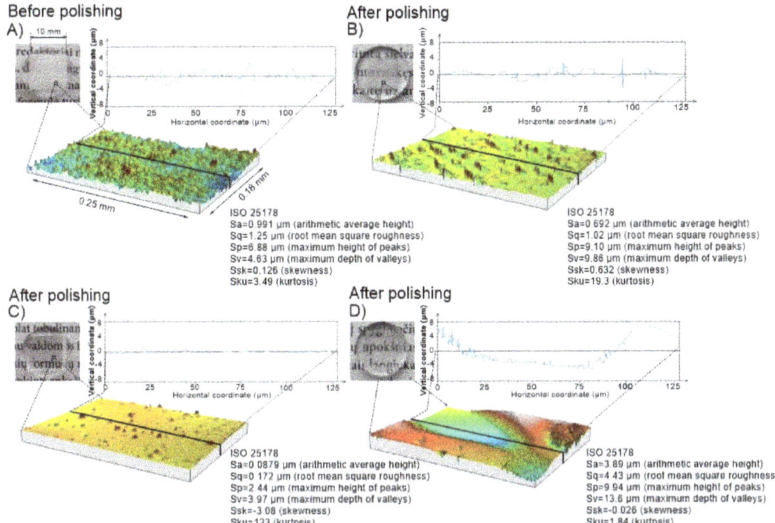

Figure 6. Visual representation of Ra (top left corner). Hydrophilic acrylic after ablation using parameters: P = 30 W, v = 3.65 m/s and dx = 36.5 μm + 2 additional scans with parameter set of P = 2.5 W, v = 1.03 m/s and dx = 10.3 μm (**A**). The same sample after polishing using the following parameters: P = 12 W, v = 1000 mm/s, dx = 5 μm and N_p = 10 (**B**); after polishing with: P = 29 W, v = 1000 mm/s, dx = 10 μm and N_p = 10 (**C**); and after polishing with P = 29 W, v = 1000 mm/s, dx = 5 μm and N_p = 10 (**D**). Sa is a different notation of Ra, notation depends on ISO standard used.

3.3. Investigation of Temperature Relation to Surface Roughness

From the previous section (see Figure 5C,D) it is evident that there are multiple micromachining parameter combinations that result in low (60.8 nm Ra) roughness values. The main cause for this is believed to be the rise in surface temperature up to a certain level to initiate melting of the rough regions. To investigate this further the temperature of the surface was recorded with an IR camera when performing the polishing procedure using different average power settings and number of sub-pulses within the burst. The 5 and 10 sub-pulses within the burst and the change of laser power from 8 to 32 W were chosen for comparison. The results are displayed in Figure 7. Correlation between the achieved surface roughness and surface temperature is evident (e.g., N_p = 5, P = 24 W and N_p = 10, P = 32 W). The measured surface roughness (57 nm and 52 nm, respectively) are excellent if maximum temperature is raised to around 357 °C. In this specific case a larger average power by two-fold is needed when using 10 sub-pulses within the burst as compared to 5 sub-pulses to achieve the best surface smoothness. This may be explained by taking into account that the absorption of light occurs via non-linear absorption processes, therefore more energy is absorbed within the material at lower average power settings when using 5 sub-pulses within the burst as compared to 10 sub-pulses, as the latter has a lower intensity. In addition, when performing the characterization of the bursts (see Figure 3) a trail of pre-pulses was registered for the 10 sub-pulse settings which was absent for the 5 sub-pulse case. How the pre-pulses affect the absorption of light and the polishing process is beyond the scope of this article, though it remains an interesting topic for future research. It was noticed that, during laser polishing when the temperature of the surface was higher than 300 °C, a high accumulated temperature zone formed (see Figure 7, residual heat) at the edge of the rectangle where the beam scanning started, in other words, the zone which is always opposite to the moving heat front and is maintained until the end of the polishing procedure. The moving temperature front is associated with the polishing beam position; as the photographs presented in Figure 7 are taken during the polishing process therefore the moving temperature front is visible. A possible explanation regarding the origins of the accumulation of temperature in the residual heat zone is the reflection of light from the slope of the laser impingement area (Figure 2, "melt pool") at the point where the molted material forms. The laser light that was reflected impinges the edge of the cuboid and increases the temperature slightly during the polishing process. Besides that, it restricts the area from cooling down via diffusion of heat. The molten material can exhibit optical quality properties and thus reflects light well. In addition, it is known [38] that a plasma cloud forms after the first pulse has impinged on the surface, and does not dissipate until the second pulse (after 400 ps) arrives at the surface. Depending on multiple machining conditions, a portion of light can also reflect from the plasma cloud. However, the stated assumptions need to be proven experimentally and remain a topic of future research.

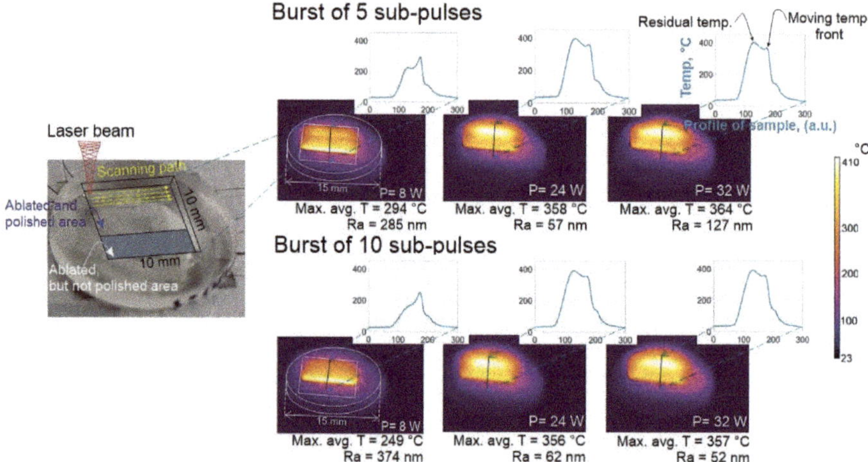

Figure 7. Temperature maps of the sample's surface during micromachining and profiles of temperature maps averaged through the working area using different average laser power for bursts of 5 and 10 sub-pulses per burst. Max. avg. T listed below the images stands for maximum averaged temperature of the moving laser front. Other parameters used for polishing were: dx = 10 μm, v = 700 mm/s and number of scans = 2. Ra of the surface are partially dependent on the position where the measurement was performed, thus errors of ±25 nm must be taken into account.

The minimal surface roughness achieved after a single scan over the surface was ~250 nm. It is known that for metals, scanning the surface multiple times increases the smoothness of the surface [39,40]. Therefore, the surface roughness was also measured when performing several scans (a change in the scanning angle by 90° is produced after each pass). It was noticed that scanning twice improves the surface finish as compared to the single scan case; however, when the scanning was done more than twice the surface roughness increases again and the sample becomes non-transparent. This may be explained by noting that after a number of scans, a deeper portion of the material reaches melting point and the material starts boiling. Bubbles rise to the surface and leave large craters as can be seen in Figure 6D. Therefore it is likely that the achieved minimal surface roughness value (Ra = 18 nm using parameters: P = 22 W, N_p = 5, v = 620 mm/s, dx = 10 μm, number of scans = 2 and f = 100 kHz) cannot be reduced further due to the Marangoni effect [41].

Figure 5 showed that by using bursts of femtosecond laser pulses it was possible to increase the surface quality. To further examine how the number of sub-pulses in the burst affects the polishing quality, we looked into how the number of sub-pulses within the burst correlates to the sample's surface temperature. Figure 8A shows that when the number of sub-pulses rises from 1 to 3, the surface temperature increases from 285 to 355 °C, while the roughness of the samples decreases to below 100 nm. Although, increase of the surface quality is only apparent until a single pulse is split into three, further increasing the number beyond this value does not increase the surface quality. These observations suggest that when splitting a single pulse into three sub-pulses, a certain energy threshold per pulse is reached and thermodynamical processes are stimulated which result in the remelting of the surface and its smoothening. Furthermore, Figure 5D shows that the temperature dependence on the average laser power is not linear. We have noticed that when increasing the average power of the laser, the temperature rises quickly to 350–360 °C. When the surface temperature increases to more than 360 °C, bubbles form below the surface. The bubbles are the release of gas trapped inside the volume which explode upon reaching the surface, this causes sputtering of the molten material and creates micro-valleys. In turn, this causes an increase in the surface roughness [42]. These experiments show, that the optimal temperature at which the best surface quality can be achieved is approximately 357 °C. The images of the surface are depicted in Figure 8B, and that no recognizable difference is seen for the case of 3–30 sub-pulses per burst.

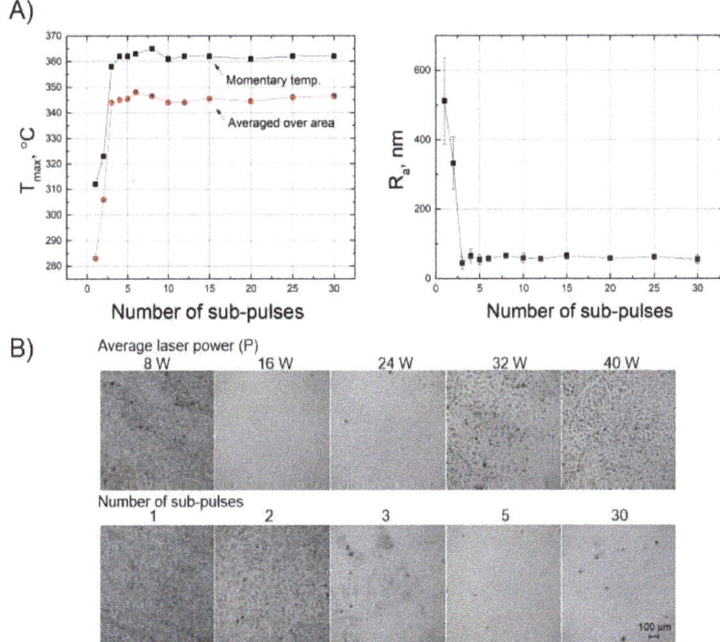

Figure 8. Surface temperature dependence on the number of sub-pulses within a burst (left picture), surface roughness dependence on the number of sub-pulses within a burst (right picture). Other parameters: P = 16 W, dx = 10 μm, v = 700 mm/s and number of scans = 2 (**A**). The surface of the sample after polishing using different average laser power when other parameters are N_p = 10, dx = 10 μm, v = 700 mm/s and number of scans = 2 (**B**, first line). The surface of the sample after polishing using different numbers of sub-pulses when other parameters are P = 16 W, dx = 10 μm, v = 700 mm/s and number of scans = 2 (**B**, second line).

Since it is thought that the Marangoni effect is the limiting factor preventing further decease in the surface roughness, experiments were performed investigating the relation between the surface roughness before polishing (prior roughness) and the surface roughness after polishing (final roughness). Initially, it was expected that larger differences would be present when performing the polishing procedure on different roughness surfaces. The experiment was carried out by preparing a number of samples, each having different initial roughness values ranging from 0.75 to 3 μm Ra. Different surface roughness values were achieved by varying the average power of the laser without using burst-mode.

The 3 µm surface roughness value was achieved by adjusting the spatial pitch of the hatching lines to dx < 5 µm. In addition, for comparison purposes, a number of samples were polished mechanically to a roughness of 40 nm Ra. After polishing the different roughness samples no significant difference in the final roughness was detected when comparing polishing with 5 and 10 sub-pulses in the burst, see Figure 9B. The results fall within an error range of one standard deviation of 50 nm. This result again confirms that as long as a certain temperature is reached on the surface during the polishing step, the results do not differ even when changing the number of sub-pulses from 5 to 20. In addition, it shows that the surfaces with roughness in range of 0.75–3 µm can be successfully polished to the same finish level using method presented in this study. However, for the case of the mechanically polished samples, further reduction of the surface roughness was not achieved. As visible in Figure 9A, small artifacts resembling dimples are present on the surface of sample after polishing with bursts, which contribute to a surface roughness value of 45 nm Ra. These artifacts are thought to be the result of embedded polishing particles in the surface, a residual from mechanical pre-polishing. Eventually, best surface quality was achieved by polishing via laser prepared samples with parameters: P = 22 W, N_p = 5, v = 620 mm/s, dx = 10 µm, f = 100 kHz and scanning twice. With the former parameter configuration we were able to achieve ~20 nm Ra and ~30 nm Sq, which according to Rayleigh roughness criteria R_{at} = 0.1 < π/16 represents a flat surface with little to zero scattering losses.

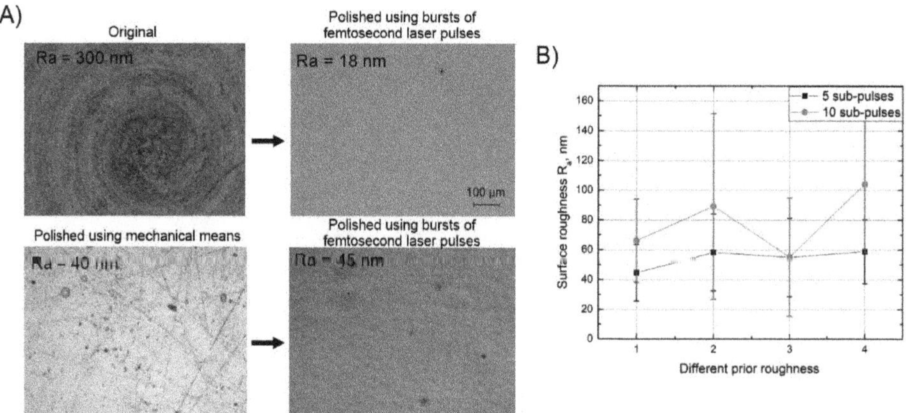

Figure 9. Pre-polished and original samples after polishing using bursts of femtosecond laser pulses (**A**), surface roughness after polishing using bursts of femtosecond laser pulses dependent on the different prior roughnesses when N_p = 5 and N_p = 20 (**B**). Parameters used in (**A**) section are: P = 22 W, N_p = 5, v = 620 mm/s, dx = 10 µm, f = 100 kHz and number of scans = 2. Parameters used in (**B**) section are: P = 32 W, v = 700 mm/s, dx = 10 µm, f = 100 kHz and number of scans = 2.

For demonstration purposes, a concave lens of 13.5 mm diameter and having a radius of curvature of 20.5 mm was fabricated. The shape was ablated using parameters: v = 2500 mm/s, P = 16 W, dx = 20 µm and polished with following parameter set: P = 22 W, N_p = 5, v = 620 mm/s, dx = 10 µm and f = 100 kHz. The result is shown in Figure 10. It is evident that the fabricated concave lens is fully transparent, smooth and bends the image like a lens.

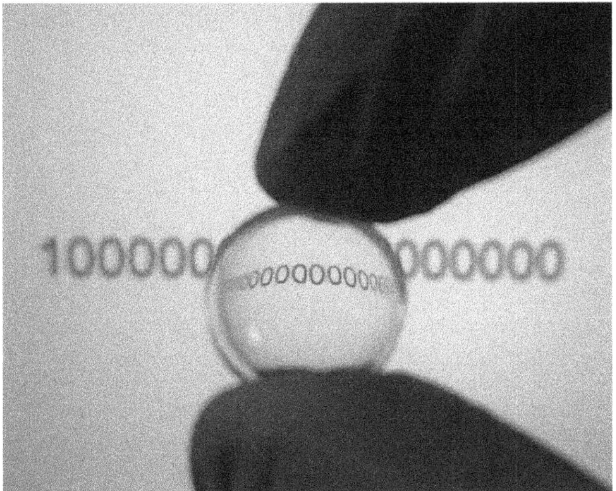

Figure 10. Polished curved surface. Polishing parameters: P = 22 W, N_p = 5, v = 620 mm/s and dx = 10 μm.

4. Conclusions

In this study we have investigated the ability to polish the hydrophilic acrylic polymer Contamac CONTAFLEX 26% UV-IOL (R), which is used as a bio-compatible polymer for body implants manufacturing, using bursts of femtosecond laser pulses. The parametric study of the laser machining parameters that consist of the following micromachining parameters: average laser power, scanning speed, scanning pitch and number of sub-pulses within the burst, showed that polishing up to a surface roughness value of Ra = 40 nm is achievable. The analysis of the surface temperature during micromachining showed that the polishing process depends on the thermal heat flux which initiates thermodynamical processes that change the surface roughness of the sample. The optimal surface temperature was found to be approximately 357 °C, at which it was possible to polish the material up to 18 nm Ra, resembling a transparent surface which meets optical-quality standards. The correlation between the temperature at the surface and the achieved surface roughness values is presented which shows that such roughness values are achievable through variation of the number of sub-pulses within the burst. In addition, it was shown that in order to reach the stated temperature levels, it is necessary to split the single pulse into at least more than three sub-pulses. However, if the number of sub-pulses is increased above three, similar results in terms of surface temperature may be achieved. If the surface is overheated to above 360 °C, micro bubbles form which rise to the surface and produce surface craters upon exploding, dramatically reducing the surface quality. It was shown that the burst-pulse polishing process is invariant of the initial roughness of the surface in the range of 0.7–3 μm Ra. In addition, if the surface has embedded mechanical polishing artifacts (abrasive particles), the surface degrades after the polishing procedure. Finally, it was shown that results achieved in this study can be applied to manufacturing of free-form optical components such as lenses, providing a fast (few minutes per lens) and efficient way of polishing to achieve low roughness surfaces.

Author Contributions: E.K. prepared the data for the article, assembled the experimental setup and organized the experiments, carried out the experiments and wrote the article. S.B. performed experiment planning, advised on technical aspects of the micromachining setup, consulted on writing of the article and reviewed and redacted the article from a professional standpoint. P.T. aided in experimental planning, quality analysis, provided valuable insight when producing 3D shapes and supplied the samples which were used for investigation. V.J. contributed to writing the article. M.B. aided in valuable technical consultation regarding the laser equipment. V.S. oversaw the entire study, provided theoretical background and consulted on writing the article. All authors have read and agreed to the published version of the manuscript.

Funding: The research was carried out in collaboration between Laser Research Center of Vilnius University and Light Conversion, UAB.

Conflicts of Interest: The authors declare no conflict of interest.

References

1. Congdon, N. Prevalence of Cataract and Pseudophakia/Aphakia among Adults in the United States. *Arch. Ophthalmol.* **2004**, *122*, 487–494. [CrossRef] [PubMed]
2. Raport ot the Joint World Health Organization-Brien Holden Vision Institute Global Scientific Meeting on Myopia. Available online: https://www.who.int/blindness/causes/MyopiaReportforWeb.pdf (accessed on 4 December 2020).
3. Leonard, P.; Rommel, J. *Lens Implantation: Thirty Years of Progress*, 1st ed.; Springer: Berlin/Heidelberg, Germany, 1982.
4. Heberle, J.; Häfner, T.; Schmidt, M. Efficient and damage-free ultrashort pulsed laser cutting of polymer intraocular lens implants. *CIRP Ann.* **2018**. [CrossRef]
5. Yu, N.; Fang, F.; Wu, B.; Zeng, L.; Cheng, Y. State of the art of intraocular lens manufacturing. *Int. J. Adv. Manuf. Technol.* **2018**. [CrossRef]
6. Schwertz, K. *An Introduction to the Optics Manufacturing Process*. OptoMechanics. 2008. Available online: https://wp.optics.arizona.edu/optomech/wp-content/uploads/sites/53/2016/10/Katie_Introduction-to-the-Optics-Manufacturing-Process.pdf (accessed on 4 December 2020).
7. Yamakawa, N.; Tanaka, T.; Shigeta, M.; Hamano, M.; Usui, M. Surface roughness of intraocular lenses and inflammatory cell adhesion to lens surfaces. *J. Cataract Refract. Surg.* **2003**. [CrossRef]
8. Valle, M.; Yamada, A.; James, R. Intraocular Lens Tumbling Rpocess Using Coated Beads. U.S. Patent 5,961,370, 5 October 1999.
9. Xia, B.; Jiang, L.; Li, X.; Yan, X.; Zhao, W.; Lu, Y. High aspect ratio, high-quality microholes in PMMA: A comparison between femtosecond laser drilling in air and in vacuum. *Appl. Phys. A Mater. Sci. Process.* **2015**. [CrossRef]
10. Heberle, J.; Klämpfl, F.; Alexeev, I.; Schmidt, M. Ultrafast laser surface structuring of intraocular lens polymers. *J. Laser Micro Nanoeng.* **2013**. [CrossRef]
11. Tonshoff, H.K.; Ostendorf, A.; Nolte, S.; Korte, F.; Bauer, T. Micro-machining using Femtosecond Lasers. In Proceedings of the First International Symposium on Laser Precision Microfabrication, Saitama, Japan, 14–16 June 2000; Volume 4088, pp. 136–139.
12. Bauer, S.; Schmuki, P.; von der Mark, K.; Park, J. Engineering biocompatible implant surfaces: Part I: Materials and surfaces. *Prog. Mater. Sci.* **2013**, *58*, 261–326. [CrossRef]
13. Louropoulou, A.; Slot, D.E.; Van der Weijden, F. Influence of mechanical instruments on the biocompatibility of titanium dental implants surfaces: A systematic review. *Clin. Oral Implant. Res.* **2015**. [CrossRef]
14. Boswald, M.; Girisch, M.; Greil, J.; Spies, T.; Stehr, K.; Krall, T.; Guggenbichler, J.P. Antimicrobial activity and biocompatibility of polyurethane and silicone catheters containing low concentrations of Silver: A new perspective in prevention of polymer-associated foreign-body-infections. *Zent. Bakteriol.* **1995**. [CrossRef]
15. Mukherjee, R.; Chaudhury, K.; Das, S.; Sengupta, S.; Biswas, P. Posterior capsular opacification and intraocular lens surface micro-roughness characteristics: An atomic force microscopy study. *Micron* **2012**. [CrossRef] [PubMed]
16. Hendow, S.; Kosa, N. Bursting improves laser material processing. *SPIE Newsroom* **2014**. [CrossRef]
17. Gaudiuso, C.; Kämmer, H.; Dreisow, F.; Ancona, A.; Tünnermann, A.; Nolte, S. Ablation of silicon with bursts of femtosecond laser pulses. In Proceedings of the Photonics West 2016—Frontiers in Ultrafast Optics: Biomedical, Scientific, and Industrial Applications XVI, San Francisco, CA, USA, 9 March 2016; Volume 9740.
18. Kerse, C.; Kalaycloĝ Lu, H.; Elahi, P.; Çetin, B.; Kesim, D.K.; Akçaalan, Ö.; Yavaş, S.; Aşlk, M.D.; Öktem, B.; Hoogland, H.; et al. Ablation-cooled material removal with ultrafast bursts of pulses. *Nature* **2016**. [CrossRef] [PubMed]
19. Knappe, R.; Haloui, H.; Seifert, A.; Weis, A.; Nebel, A. Scaling ablation rates for picosecond lasers using burst micromachining. In Proceedings of the SPIE LASE, Laser-based Micro- and Nanopackaging and Assembly IV, San Francisco, CA, USA, 23 February 2010; Volume 7585.

20. Mayerhofer, R. Ultrashort-pulsed laser material processing with high repetition rate burst pulses. In Proceedings of the SPIE LASE, Laser Applications in Microelectronic and Optoelectronic Manufacturing (LAMOM) XXII, San Francisco, CA, USA, 20 February 2017; Volume 10091.

21. Butkus, S.; Jukna, V.; Paipulas, D.; Barkauskas, M.; Sirutkaitis, V. Micromachining of invar foils with GHz, MHz and kHz femtosecond burst modes. *Micromachines* **2020**, *11*, 733. [CrossRef] [PubMed]

22. Žemaitis, A.; Gečys, P.; Barkauskas, M.; Račiukaitis, G.; Gedvilas, M. Highly-efficient laser ablation of copper by bursts of ultrashort tuneable (fs-ps) pulses. *Sci. Rep.* **2019**. [CrossRef]

23. Barkauskas, M.; Neimontas, K.; Butkus, V. Device and Method for Generation of High Repetition Rate Laser Pulse Bursts. U.S. Patent 2020/0067260 A1, 27 February 2020.

24. Becker, W. *Advanced Time-Correlated Single Photon Counting Techniques*, 1st ed.; Springer: Berlin/Heidelberg, Germany, 2005.

25. Bonse, J.; Rudolph, P.; Krüger, J.; Baudach, S.; Kautek, W. Femtosecond pulse laser processing of TiN on silicon. *Appl. Surf. Sci.* **2000**. [CrossRef]

26. Temmler, A.; Willenborg, E.; Wissenbach, K. Laser Polishing. In Proceedings of the SPIE LASE, Laser Applications in Microelectronic and Optoelectronic Manufacturing (LAMOM) XVII, San Francisco, CA, USA, 15 February 2012; Volume 8243.

27. Perry, T.L.; Werschmoeller, D.; Li, X.; Pfefferkorn, F.E.; Duffie, N.A. Pulsed laser micro polishing of microfabricated nickel and Ti6Al4V samples. In Proceedings of the ASME International Manufacturing Science and Engineering Conference, MSEC2008, Evanston, IL, USA, 7–10 October 2008.

28. Temmler, A.; Willenborg, E.; Wissenbach, K. Design surfaces by laser remelting. In Proceedings of the Sixth International WLT Conference on Lasers in Manufacturing, Munich, Germany, 23–26 May 2011.

29. Nüsser, C.; Wehrmann, I.; Willenborg, E. Influence of intensity distribution and pulse duration on laser micro polishing. In Proceedings of the Sixth International WLT Conference on Lasers in Manufacturing, Munich, Germany, 23–26 May 2011.

30. Mai, T.A.; Lim, G.C. Micromelting and its effects on surface topography and properties in laser polishing of stainless steel. *J. Laser Appl.* **2004**. [CrossRef]

31. Temmler, A.; Graichen, K.; Donath, J. Laser Polishing in Medical Engineering. *Laser Tech. J.* **2010**. [CrossRef]

32. Reeber, R.R. Surface Engineering of Structural Ceramics. *J. Am. Ceram. Soc.* **1993**. [CrossRef]

33. Bereznai, M.; Pelsöczi, I.; Tóth, Z.; Turzó, K.; Radnai, M.; Bor, Z.; Fazekas, A. Surface modifications induced by ns and sub-ps excimer laser pulses on titanium implant material. *Biomaterials* **2003**. [CrossRef]

34. Gadelmawla, E.S.; Koura, M.M.; Maksoud, T.M.A.; Elewa, I.M.; Soliman, H.H. Roughness parameters. *J. Mater. Process. Technol.* **2002**. [CrossRef]

35. Nam, J.R.; Kim, C.H.; Jeoung, S.C.; Lim, K.S.; Kim, H.M.; Jeon, S.J.; Cho, B.R. Measurement of two-photon absorption coefficient of dye molecules doped in polymer thin films based on ultrafast laser ablation. *Chem. Phys. Lett.* **2006**. [CrossRef]

36. Baudach, S.; Bonse, J.; Krüger, J.; Kautek, W. Ultrashort pulse laser ablation of polycarbonate and polymethylmethacrylate. *Appl. Surf. Sci.* **2000**. [CrossRef]

37. Pinel, N.; Bourlier, C.; Saillard, J. Degree of roughness of rough layers: Extensions of the rayleigh roughness criterion and some applications. *Prog. Electromagn. Res. B* **2010**. [CrossRef]

38. Hernandez-Rueda, J.; Puerto, D.; Siegel, J.; Galvan-Sosa, M.; Solis, J. Plasma dynamics and structural modifications induced by femtosecond laser pulses in quartz. *Appl. Surf. Sci.* **2012**, *258*, 9389–9393. [CrossRef]

39. Metzner, D.; Lickschat, P.; Weißmantel, S. High-quality surface treatment using GHz burst mode with tunable ultrashort pulses. *Appl. Surf. Sci.* **2020**. [CrossRef]

40. Metzner, D.; Lickschat, P.; Weißmantel, S. Influence of heat accumulation during laser micromachining of CoCrMo alloy with ultrashort pulses in burst mode. *Appl. Phys. A Mater. Sci. Process.* **2020**. [CrossRef]

41. Pfefferkorn, F.E.; Duffie, N.A.; Li, X.; Vadali, M.; Ma, C. Improving surface finish in pulsed laser micro polishing using thermocapillary flow. *CIRP Ann. Manuf. Technol.* **2013**. [CrossRef]

42. Huang, Y.; Liu, S.; Yang, W.; Yu, C. Surface roughness analysis and improvement of PMMA-based microfluidic chip chambers by CO 2 laser cutting. *Appl. Surf. Sci.* **2010**. [CrossRef]

Publisher's Note: MDPI stays neutral with regard to jurisdictional claims in published maps and institutional affiliations.

 *micromachines*

Article

Functionalization of Plastic Parts by Replication of Variable Pitch Laser-Induced Periodic Surface Structures

Leonardo Piccolo [1,2], **Marco Sorgato** [2], **Afif Batal** [3], **Stefan Dimov** [3], **Giovanni Lucchetta** [2] and **Davide Masato** [1,*]

1   Department of Plastics Engineering, University of Massachusetts Lowell, Lowell, MA 01854, USA;
    Leonardo.piccolo@phd.unipd.it
2   Department of Industrial Engineering, University of Padova, 35131 Padova, Italy;
    Marco.sorgato@unipd.it (M.S.); giovanni.lucchetta@unipd.it (G.L.)
3   Department of Mechanical Engineering, University of Birmingham, Birmingham B15 2TT, UK;
    bxa361@student.bham.ac.uk (A.B.); s.s.dimov@bham.ac.uk (S.D.)
*   Correspondence: Davide_Masato@uml.edu; Tel.: +1-(978)-934-2836

Received: 3 April 2020; Accepted: 18 April 2020; Published: 20 April 2020

Abstract: Surface functionalization of plastic parts has been studied and developed for several applications. However, demand for the development of reliable and profitable manufacturing strategies is still high. Here we develop and characterize a new process chain for the versatile and cost-effective production of sub-micron textured plastic parts using laser ablation. The study includes the generation of different sub-micron structures on the surface of a mold using femtosecond laser ablation and vario-thermal micro-injection molding. The manufactured parts and their surfaces are characterized in consideration of polymer replication and wetting behavior. The results of the static contact angle measurements show that replicated Laser-Induced Periodic Surface Structures (LIPSSs) always increase the hydrophobicity of plastic parts. A maximum contact angle increase of 20% was found by optimizing the manufacturing thermal boundary conditions. The wetting behavior is linked to the transition from a Wenzel to Cassie–Baxter state, and is crucial in optimizing the injection molding cycle time.

Keywords: laser-induced periodical surface structures; micro-injection molding; replication; surface wettability

---

1. Introduction

Polymers are materials characterized by very different properties that can be tailored for many applications [1]. Engineers are constantly trying to enhance their desired properties or to straighten the weaknesses of these materials. Surface functionalization has emerged as a solution to the needs of various stakeholders. Functional surfaces find application as self-cleaning [2–4], scaffolds [5,6], tissues [7–9], optical parts [10], anti-icing [11,12], in friction reduction [13,14], and in engineering of injection mold surfaces [15]. All these applications generated an increasing demand to develop reliable and cost-effective mass manufacturing technologies for polymer surface replication.

Functionalization of polymer surfaces can be obtained by modifying their chemistry (e.g., using coatings) or through topographic modifications (e.g., texturing) [16]. The chemical modification of a surface typically requires an additional step on the production of the part. This paper focuses on tailoring the surface properties of plastic parts through topographic modification, exploiting a net shape process. Functionalization can be described by considering the wettability of a surface [17]. Wettability is a fundamental characteristic of solids, and is governed by surface chemistry and morphology. It measures the ability of a liquid to maintain contact with a solid surface, and is determined by the balance between cohesive (inside the liquid) and adhesive (between liquid and the solid surface) forces. By changing the surface topography, adhesive interactions can be modified. Wettability can be characterized by utilizing contact angle measurements. A hydrophilic surface exhibits a contact angle smaller than 90°. Conversely, a hydrophobic surface yields a contact angle above 90°. When the contact angle exceeds 150°, the surface is defined as super-hydrophobic.

When considering surface topographies, two models were developed to explain the interaction between asperities and liquid droplets. The Wenzel state assumes that liquid droplets follow the asperities of the surface, emphasizing the intrinsic hydrophobic or hydrophilic behavior of the material [18]. The Cassie–Baxter state predicts that asperities can trap air bubbles inside them, resulting in an inhomogeneous surface [19]. When the fraction of a solid in contact with a water droplet decrease, it means that more air is trapped, and thus the contact angle increases. The Wenzel state depends directly on surface roughness, the Cassie–Baxter state depends on the wet fraction of the solid. The aim of surface structuring is to create patterns that optimize surface roughness and minimize the solid fraction of a surface in contact with the liquid.

Texturing of plastic parts can be achieved either via direct patterning or by replicating the surface of a tool. The former approach is slow, expensive and may lead to some chemical changes on the surface [20,21], while the latter offers more efficient manufacturing opportunities. Indeed, the high investment costs of patterning the mold can be adequately justified when the tool is used for long production runs. Mold texturing technologies can be divided into a top-down (i.e., removing material) and a bottom-up (adding material) approach [22]. Top-down technologies are approaches that shape the pattern by removing material from the substrate using either hard tools (e.g., milling, etching) or energy (e.g., laser engraving, electrical discharge machining). Bottom-up technologies exploit self-assembly or self-organization properties of certain materials to create small (i.e., up to few nanometers) and accurate structures. The investment and operating costs of top-down techniques are high, and there are substantial limitations on the patterning of large areas and complex surfaces. At the same time, despite their easier scalability, bottom-up technologies are characterized by the worst capabilities in terms of geometry, and feature size is limited [23].

Mold texturing for injection molding applications requires a high speed and flexibility [24]. Laser patterning has been proven effective for the fast-engraving of highly uniform micro-structured surfaces on a large variety of materials, including steel [25–27]. Furthermore, throughputs as high as 1 cm^2 per 10 s have been reported for metals when producing self-organized sub-micron structures [28]. High texturing speeds characterize this type of surface modification due to the unique structuring phenomena happening on the surface. The laser beam irradiates the surface on an area that is much bigger than the feature size, leaving a sort of fingerprint characteristic of the particular laser beam. Femtosecond lasers, typically exploited for the formation of Laser-Induced Periodic Surface Structures (LIPSSs), are pulsed lasers characterized by extremely short pulse durations and high peak power. The interference pattern formed by the incident wave (i.e., the laser light wave) and the exited Surface Electromagnetic Wave (SEW) rule the patterning phenomena. The repeated interference between the two waves creates a pattern of energy distribution that ionizes and melts the very first layer of material on the surface, accordingly creating the typical ripples of the LIPSS pattern [29].

The morphology of the LIPSS pattern can be modified by changing the laser setup. The regularity and depth of the structures can be optimized, controlling the way the beam scans the surface and the beam fluence (i.e., energy delivered on a specific area). The pitch between consecutive structures is strongly linked to the wavelength of the laser light (typically, the pitch is slightly smaller than the laser wavelength [30]).

Tunable laser systems (variable beam wavelength) are more expensive than their single-frequency counterparts due to the added complexity, while their output power is generally compromised. Therefore, other approaches can be considered to modify LIPSS periodicity. Varying the laser Beam Incident Angle (*BIA*) relative to the surface can be one method for varying the ripples' pitch. The phenomenon is controlled by the energy interference pattern formed by the incident wave and the exited Surface Electromagnetic Wave (SEW). This interference may result in a different period of the pattern or a pattern with two periods. For a Transverse Magnetic (TM) wave, the period of the structures is given by

$$\Lambda_{1,2} = \frac{\lambda}{1 \pm \sin \theta} \tag{1}$$

where Λ is the pattern pitch, λ is the laser wavelength and θ the angle between the laser beam and the normal surface direction [31].

Hot embossing and injection molding are the most efficient technologies to achieve replication of submicron-structured plastic parts [32]. However, when it is required to apply textures on 3D geometries, only micro injection molding (µIM) can be considered [33]. The replication of submicron-structured surfaces is challenging since the injected molten polymer tends to solidify quickly touching the cold mold, hindering filling of the micro-cavities [34]. Replication is highly influenced by process parameters (mold temperature, melt temperature, injection velocity and holding pressure) and mold design (e.g., distance of the patterned area from the gate and part thickness) [34,35]. In the literature, many experimental studies have focused on the optimization of the µIM process and auxiliaries to overcome the manufacturing issues driven by the replication of high aspect ratio structures [36,37]. Rapid Heat Cycle Molding (RHCM) is commonly used to increase mold temperatures above the polymer glass transition temperature (T_g) during the injection phase. The fast cooling of the mold after filling allows the ejection of the part [38]. However, RHCM requires significant investments, and the cycle time is highly affected.

The replication performance of the process is generally evaluated in terms of surface topography without considering its functionality. In this work, the functionalization of the surface is characterized considering its wetting properties. The correlation between replication and wettability is not clear yet. With functionalization of the surface being the aim of the whole process, understanding the roles of the process features on the wetting phenomena can drive process optimization and productivity.

In this work, the process chain for manufacturing functional submicron-structured plastic surfaces is studied. A femtosecond laser source is used to generate different submicron structures on the mold surface. The laser ablation process is characterized and studied to understand the effect of its parameters on ripple morphology. The mold surface textures are then replicated by the injection molding of different thermoplastic resins. Texture design, laser processing and the wettability of parts are studied holistically to optimize the manufacturing process as well as the functionality of the parts.

2. Materials and Methods

2.1. Process Chain and Experimental Approach

In this work, the process for cost-effective and fast production of functionalized plastic parts using a femtosecond laser is presented and characterized. The proposed process chain accomplishes the plastic part functionalization by laser texturing of the mold surface and replication by µIM (Figure 1). The sub-micrometric pattern on the conventional steel surface is engraved using an ultrafast pulsed laser. The texture replication is accomplished using µIM, since it is a suitable process to replicate even small features on 3D parts with a wide range of thermoplastic materials at a low cycle time. Texturing of plastic parts leads to improved wetting properties.

The process chain was validated by texturing four mold inserts with different laser process conditions. The µIM process capability to replicate the surface structures was experimentally investigated for different temperature boundary conditions. As the mold surface temperature decreased, the melt cooling rate increased, leading to higher viscosities and preventing the replication of the submicron structures. The process performance was evaluated considering the functionality of the manufactured plastic parts. The correlation between surface functionalization and polymer replication is discussed comparing the plastic surface wetting behavior with the replicated topographies.

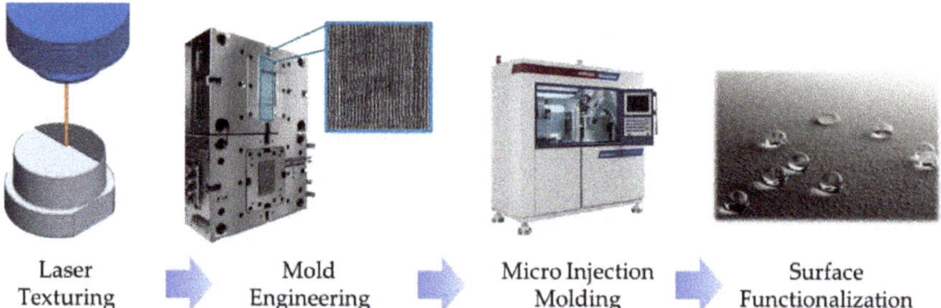

Figure 1. Schematic of the process chain utilized in this work to functionalize plastic parts.

2.2. Mold Design

The plastic part designed for this study was a disk with a diameter of 18 mm and a wall thickness of 1.5 mm. The cavity was placed on the B-side of the mold at the end of a 7-mm long semi-circular cold runner with a radius of 1.5 mm (Figure 2a). The textured surface had a diameter of 10 mm and was assembled in the cavity using round inserts (Figure 2b). Two alignment slots were machined on the side of the insert. They were used to align the surface structures with the polymer flow. Ejector pins were evenly distributed around the cavity to ensure uniform demolding force on the molded plastic part. Electrical heating was combined with water cooling for the RHCM process and to control the mold surface temperature.

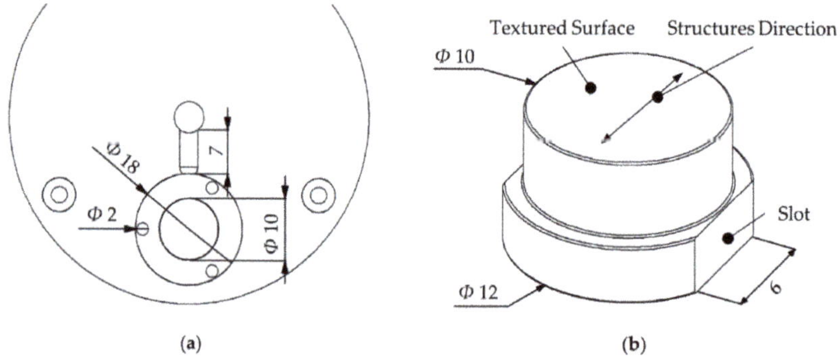

Figure 2. (a) Design of the mold cavity. (b) Design of the textured insert and alignment features. (All dimensions are in millimeters.)

2.3. Mold Surface Treatments

Five mold inserts were machined and polished to obtain a smooth surface finish (Sa below 25 nm) before the generation of the submicron structures by laser ablation. Then, the surface of the inserts was treated with ultrashort laser patterning using a femtosecond laser source (Satsuma, Amplitude Systems, Pessac, France). The laser process parameters, reported in Table 1, were selected based on a previous optimization study to obtain a uniform pitch and depth of the ripples [39]. The laser beam polarization was unchanged during the trials, and the inserts were oriented identically in the Computer Numerical Control (CNC) machining setup, ensuring the same ripples' directions over the four textured samples. The irradiation was done in such a way ensuring the beam p-type linear polarization was perpendicular to the scanning direction. Laser patterning was carried out by focusing the laser beam using a telecentric lens down to a diameter ($2w_0$) of 40 µm and moving it with a galvanometer scanner over the insert surface at a speed of (v) 1500 mm/s with a pulse frequency (f) of 250 kHz. Thus, the pulse distance (L) was 6 µm (distance between two consecutive pulses). The same value was used for the scanning step over distance (*H*) (spacing between two consecutive scanning lines), thus obtaining the same pulse overlaps in both X and Y directions.

Table 1. Laser process parameters.

Parameter	Value
Wavelength (nm)	1030
Pulse duration (fs)	310
Pulse energy (μJ)	2.52
Frequency (*f*) (kHz)	250
Scanning speed (*v*) (mm/s)	1500
Step over (*H*) (μm)	6
Spot diameter ($2w_0$) (μm)	40
Fluence (*F*) at normal incidence (mJ/cm^2)	200

The pitch distance between consecutive ripples was changed in different inserts by changing the beam incidence angle (*BIA*). *BIA* values were changed between 0° (normal incidence) and 30° while keeping the polarization parallel to the plane of incidence. In particular, the four inserts A, B, C and D were produced with *BIAs* of 0°, 10°, 20°, and 30°, respectively. The pulse fluence is dependent on the beam profile on the insert surface, and therefore it is slightly reduced as the beam area increases when the *BIA* value is increased. When the surface is tilted the beam spot can be assumed to be an ellipse with the axis size a trigonometric function of *BIA* [40]. Thus, the respective pulse fluence (*F*) values for *BIAs* 10°, 20° and 30° are 197, 188 and 174 mJ/cm^2, respectively.

2.4. Surface Characterization

The roughness of the polished inserts was measured before laser patterning using a confocal profilometer (3D S Neox, Sensofar, Barcelona, Spain). The instrument was used in confocal mode with a 100X lens, and surface roughness (*Sa*) of the inserts was evaluated.

The morphology of the sub-microstructures on the mold insert was characterized using Scanning Electron Microscopy (SEM - FEI, Quanta 400, Thermo Fisher Scientific, Hillsboro, OR, USA). This technique permits the homogeneity of the texture of the entire surface to be quickly examined.

The geometrical characterization of the ripples was carried out using Atomic Force Microscopy (AFM –CP II, Veeco Digital Instruments, Bruker, Billerica, MA, USA). This first morphological characterization is needed to prove the representativeness of the small AFM scanned area. The AFM analysis (Table 2) was repeated onto two 10- × 10-μm areas over each insert. The mean height of the structures was evaluated with the analysis of the acquired point clouds exploiting the Abbott-Firestone curves method according to the standard DS/EN ISO 25178-2 [41]. The material ratio curve (i.e., Abbott-Firestone curve) describes the increase of the material portion of the surface with increasing the roughness depth. For each acquired topography, the core surface roughness was determined from the linear representation of the material ratio curve and addressed to the mean topographic height.

Table 2. AFM measuring setup for the topography of the plastic parts.

Property	Insert	Part
Tip material	Silicon	Gold
Tip curvature radius (nm)	6	10
Cantilever length (μm)	180	125
Cantilever width (μm)	18	30
Cantilever thickness (μm)	0.6	3
Force constant (N/m)	0.05	1.45
Resonant frequency (kHz)	22	87

The structure periodicity was carried out using 2D FFT (2D Fast Fourier Transform) analysis of the entire measured surface. The main peaks of the spectrum can be correlated to surface periodicities [42].

Two parameters were considered to describe the pattern quality, (i) regularity (R), and (ii) homogeneity. Pattern regularity refers to the uniformity of the pitch distance between consecutive structures. This parameter was evaluated from the 2D FFT analysis of the topographies acquired using AFM. The regularity is represented by the standard deviation of the normal distribution associated with the peak of the considered ripple periodicity. Since the pitch variation was within the direction of the ripples, a horizontal profile of the 2D FFT analysis was extracted (Figure 3b). The regularity of the pattern was evaluated considering the ratio between the standard deviation (σ) of the normal distribution associated with each pitch (Λ) present along with the pattern and the pitch itself:

$$R = 1 - \sum_{i=1}^{n} \frac{2\sigma_i}{\Lambda_i} \qquad (2)$$

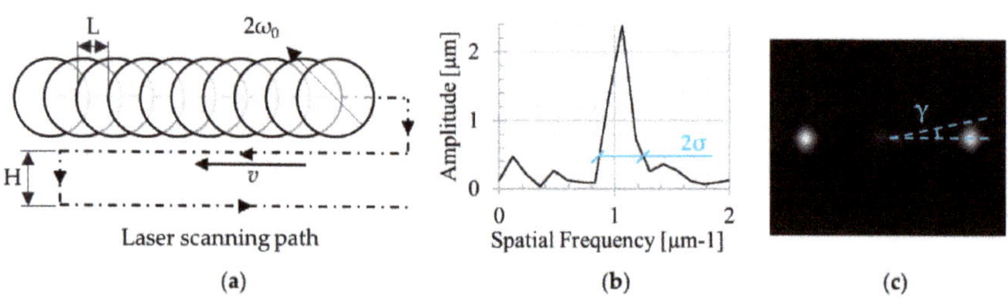

Figure 3. (**a**) Scanning path for the laser surface structuring (clarity dimensions are not correctly scaled); (**b**) FWMH (Full Width at Half Maximum) evaluation by the horizontal profile extracted from the 2D FFT of the AFM point clouds; (**c**) representative 2D FFT micrographs describing the dispersion angle.

The homogeneity of the texture was analyzed considering the ripple dispersion angle γ (i.e., the angle deviation of the ripples from the straight parallel configuration). This dispersion angle can be adequately described considering the angle γ that encloses the areal 2D FFT peak dispersion of the ripple frequency (Figure 3c).

The same morphologic and topographic analyses were performed on the plastic part surfaces to investigate the replication accuracy. SEM analysis was carried to examine the homogeneity of the replication over the entire surface. The AFM setup was adapted for the measurements on the plastic parts to optimize accuracy (Table 2).

2.5. Materials and Manufacturing System

A commercial polystyrene (PS Crystal 1540, Total Petrochemicals & Refining, Bruxelles, Belgium) and a poly-methyl methacrylate (PMMA, Polycasa Acryl G55, 3A Composites, Statesville, NC, USA) were selected for the experimental work. Their main properties are reported in Table 3. These polymers are commonly used in µIM due to their high flowability, high optical clarity, high transparency, and high impact strength.

A state-of-the-art micro injection molding machine (MicroPower 15, Wittmann-Battenfeld GmbH, Kottingbrunn, Austria) was used for the µIM experiments. The accuracy and repeatability of the injection phase were guaranteed by the separation of the metering (14 mm screw) and injection (5 mm plunger) units. A maximum injection pressure of 2700 bar was used as the limit during the molding cycles. The mold temperature was varied from 40 to 120 °C, and a 10 °C gap was chosen between the tested temperatures. The other process parameters of the injection molding for each material were kept constant (Table 4). To ensure the stability of the process, 20 molding cycles were performed before the first part was collected. For each condition, 15 parts were sampled.

Table 3. Main properties of the selected polymers.

Property	PS	PMMA
Density (g/cm^3)	1.05	1.18
Melt flow index (g/10 min)	12 (200 °C-5 kg)	21 (230 °C-3.80 kg)
Glass transition temperature (°C)	100	105
Drying time (h)	2	6
Drying temperature (°C)	70	80
Recommended ejection temperature (°C)	85	85
Recommended mold temperature range (°C)	30–50	40–90

Table 4. Main processing conditions chosen for the injection molding experiments with the two resins.

Parameter	PS	PMMA
Melt temperature (°C)	235	255
Back pressure (MPa)	5	2
Switch-over pressure (MPa)	80	82
Injection speed (mm/s)	110	110
Packing pressure (MPa)	45	50
Cooling time (s)	10	10
Mold temperature range (°C)	40–120	40–120

2.6. Wetting Characterization

The wetting properties of the textured plastic surface were evaluated from contact angle measurements and benchmarked against those of a smooth plastic surface. The equipment used for the measurements consisted of a horizontal stage, a motor-driven micrometer syringe (UMP3) with a needle (diameter 0.21 mm), two background illumination sources (LED Pholox) and two cameras (MANTA G-146, Allied Vision Technologies GmbH, Exton, PA, USA) with a telecentric 2X lens (VS-TC2-110, VS Technology, USA) (Figure 4a). The measurements were carried out using water droplets with a total volume of 500 nL.

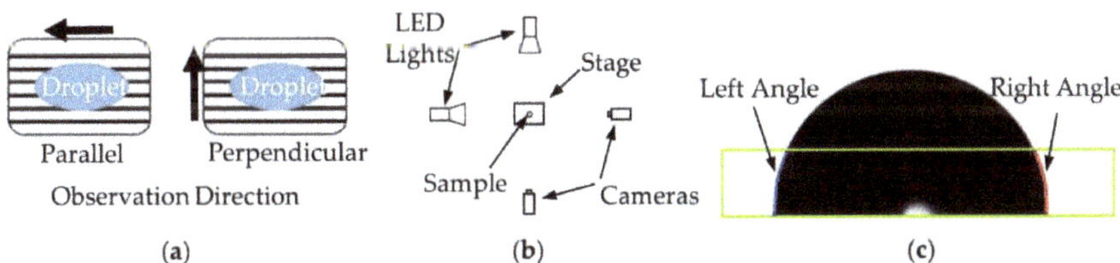

Figure 4. (**a**) Observation direction of the water droplets on the textures surface; (**b**) layout of the wetting test setup; (**c**) estimation of the contact angles through the fitting of the droplet shape. The green square defines the searching area that the software keeps.

The contact angle was measured along two directions at 90° to account for the effect of ripple directionality on the drop shape (Figure 4b). The acquired images were elaborated by fitting the shape of each droplet to calculate the contact angle, as shown in Figure 4c. The wetting characterization was performed on three parts made with each polymer for each combination of processing parameters. On each plastic part, six water droplets were deposited and their contact angles acquired. This allowed for the acquisition of 36 contact angle measurements per experimental condition.

3. Results

3.1. Generated Mold Topographies

Laser patterning of the 10-mm diameter circle required about 25 s, leading to a surface patterning speed of about 3.3 mm²/s. This demonstrates the speed of the ultrafast laser technology and its flexibility. Indeed, scaling up the texturing process would be possible, and processing times would be reasonable, allowing elevated texturing speeds and customization.

The morphology in terms of homogeneity and regularity of the sub-micrometric structures was initially characterized using SEM. SEM micrographs indicate that the LIPSSs were uniform over the entire insert surface, confirming the reliability of the laser processing at high scan speeds. Considering the SEM pictures in Figure 5, changes in the ripples' morphology could be observed when increasing the *BIA* value. In particular, the ripples' pitch and overall regularity appeared to decrease as the *BIA* increased. The periodicity of the LIPSSs was varied by changing the angle of incidence, and a second period was created in the pattern. The appearance of a second period proves that beam tilting is an effective method to change the surface texture morphology.

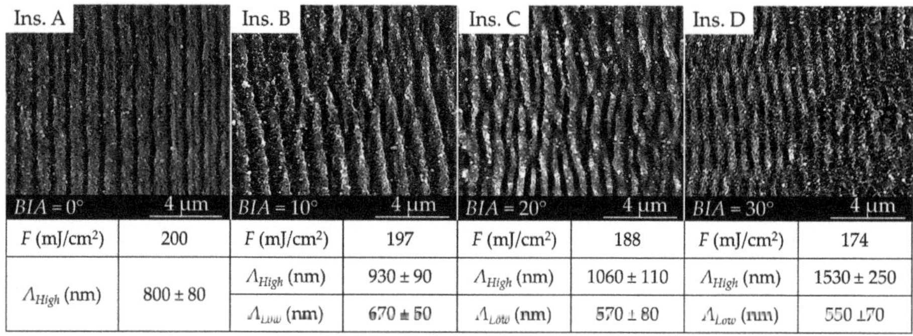

$BIA = 0°$	4 µm	$BIA = 10°$	4 µm	$BIA = 20°$	4 µm	$BIA = 30°$	4 µm
F (mJ/cm²)	200	F (mJ/cm²)	197	F (mJ/cm²)	188	F (mJ/cm²)	174
Λ_{High} (nm)	800 ± 80	Λ_{High} (nm)	930 ± 90	Λ_{High} (nm)	1060 ± 110	Λ_{High} (nm)	1530 ± 250
		Λ_{Low} (nm)	670 ± 50	Λ_{Low} (nm)	570 ± 80	Λ_{Low} (nm)	550 ± 70

Figure 5. SEM micrographs of the textured surfaces on the steel mold inserts for different beam incidence angles (*BIAs*).

The AFM analysis allowed a quantitative characterization of the inserts' topographies; in particular, the pitch distance between consecutive ripples and the homogeneity of the pattern were evaluated. These parameters were studied by considering the laser beam incidence angle and by defining and plotting some topographical parameters.

SEM micrographs suggested that changes in *BIA* could lead to the onset of a secondary periodicity in the ripples pattern. Here, we call Λ_{high} the higher (primary) pitch distance and Λ_{low} the smaller (secondary) pitch distance. The pitch distance for the structures can be calculated for different tilting angles as follows:

$$\Lambda_{Low,High} = \frac{\varepsilon \cdot \lambda}{1 \pm \sin(BIA)} \tag{3}$$

where λ is the laser wavelength, and ε is a loss factor defined to consider the wavelength-to-pitch ratio. In fact, when the beam was normalized to the workpiece surface (i.e., $BIA = 0°$), a pitch of 770 nm was measured, which is slightly lower than the laser wavelength (i.e., 1030 nm). To account for this phenomenon, a loss factor $\varepsilon = \Lambda_\alpha = 90°/\lambda$ was introduced in the theoretical model [31]. The experimental data are plotted in Figure 6a against the model defined in Equation (3 to further understand the ripple periodicity. It can be observed that the difference between the two periods became larger as the *BIA* increased. This suggests the possibility of modifying surface functionality and potentially generating hierarchic textures. The influence of *BIA* on structures' aspect ratios has not yet reported in the literature. However, an initial increase then decrease of the aspect ratio was observed with the *BIA* increased, as shown in Figure 6a. Figure 6b shows the effect of *BIAs* on ripple quality and homogeneity. It was observed that pattern regularity decreases as the *BIA* increases. This is probably related to the interference phenomena occurring on the surface during the short pulse irradiation durations, and the formation of irregular intermediary ripples in-between the primary periodic structures. This trend has also been confirmed by the dispersion angle γ.

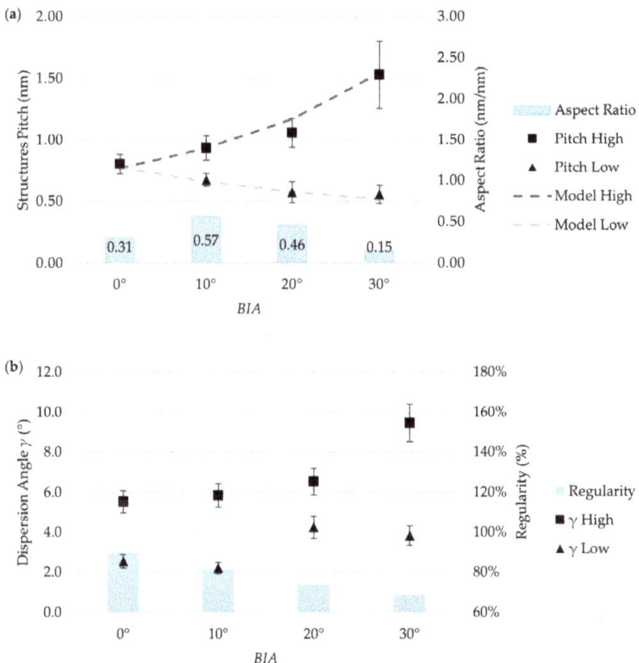

Figure 6. (**a**) Results of the AFM characterization of the four laser-machined steel mold inserts in terms of pitch and aspect ratio of the pattern; (**b**) plot of the pattern homogeneity study results for the four mold inserts.

The effects of *BIA* on both pattern regularity and homogeneity can also be evaluated using SEM micrographs (Figure 5). The pattern regularity describes how consistent the ripples' periodicity was, whereas the dispersion angle was linked to their propagation direction. Variations in pattern periodicity within a sample translated to a decrease in *R*, while the formation of intermediary ripples that bifurcated and intertwined resulting in higher dispersion angle γ (Figure 6b).

Considering the results of the metrological characterization carried out on the different textures, Insert B was selected for the injection molding experimental campaign. Indeed, at *BIA* = 10°, the ripples presented the best compromise, exhibiting both a high aspect ratio and homogeneity while displaying a low dispersion angle value. Insert B was selected for the polymer replication studies because its topography was characterized by the highest aspect ratio, which is well-known as the key parameter to controlling the replication of textured surfaces [43].

3.2. Sub-Micron Structures Polymer Replication

The surface of the molded plastic parts was initially characterized using SEM. The SEM analysis showed that the quality of the replicated ripples was homogenous along the textured surface for both polymers and for different mold temperatures (Figure 7). Micrographs show the absence of significant defects or areas that are not homogeneously replicated.

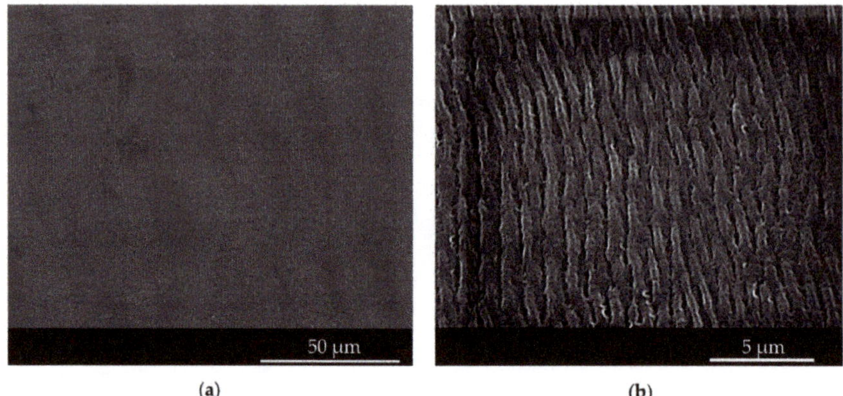

Figure 7. SEM images of the ripples replicated on PS samples. (**a**) Low magnification SEM picture of a PS sample molded at a mold temperature of 50 °C; (**b**) high magnification pictures of PS molded at 50 °C.

The AFM scans were processed using the Abbott-Firestone curves method. Core height (S_k) was evaluated for the different topographies and compared to mold measurements. Figure 8a shows the replication performance of the two resins, indicating the difference between the replicated aspect ratio and that of the mold. Moreover, the effect of mold temperature was analyzed, indicating that, as expected, higher mold temperatures lead to higher replications. However, the effect of mold temperature was different for the two resins. When molding PMMA, the effect of increasing mold surface temperature was more significant than for PS. This could be related to the temperature dependency of PMMA polymer melt viscosity. In fact, as the hot melt touches the cold mold, its viscosity increases significantly. This viscosity rises strongly depending on the polymer/mold thermal boundary and polymer properties. Figure 8b compares the Newtonian viscosity for the two polymers as a function of temperature. The behavior of the two polymers intersects around 210 °C, which is below melt temperature. Thus, during molding, PMMA has a higher melt viscosity than PS. This explains the smaller replication obtained when using PMMA and the higher sensitivity to an increase of mold temperature.

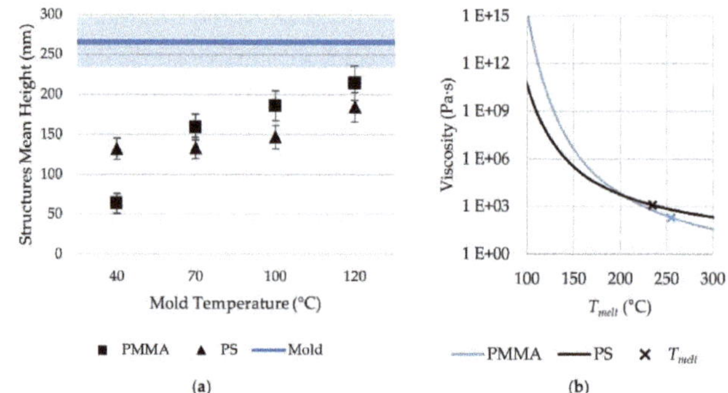

Figure 8. (**a**) Replicated structure heights for PMMA and PS. (**b**) Zero shear viscosity dependence on temperature for PMMA and PS.

3.3. Plastic Parts Surface Wettability

The analysis of the contact angles indicates that the wetting properties of plastic parts were affected by the replicated sub-micron structures. Due to the anisotropy of the textured surface, the drop has an ellipsoidal shape. To account for this effect, the contact angle along two directions (see Figure 4) was measured. The results reported in Figure 9 show that the parallel angle assumed higher values than the perpendicular one, indicating that water drops spread easier along the ripples. The maximum water contact angle increase was about 20% for PMMA and 17% for PS with respect to the smooth part.

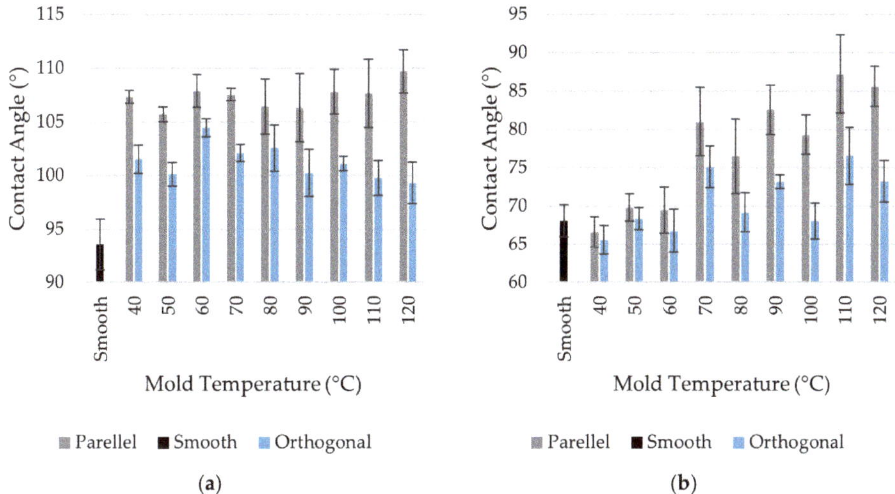

Figure 9. Results of contact angle measurements for (a) PS and (b) PMMA plastic parts.

The two polymers showed different wetting properties as a function of the mold temperature. Indeed, molding PMMA at a higher temperature substantially improved the functionalization of the surface. In particular, the most significant contact angle increase (i.e., 17%) was observed when increasing the temperature from 60 to 70 °C. This allowed the identification of a threshold value above which increasing the mold temperature did not yield significant variations in part functionality. Conversely, the wetting properties of the PS molded part did not indicate any substantial effect of the mold temperature in the investigated range.

3.4. Comparison between Achieved Replication and Functionalization

The plot shown in Figure 10 compares the polymer replication grade of submicron structures and the functionalization grade. The replication grade expresses the ratio between the height of the structures on the polymer part and the depth of the mold structures. The functionalization grade describes the contact angle gain for the textured plastic parts with respect to the smooth ones. For parts molded with PS, the grade of functionalization and replication follows the same trend, showing similar results for all tested mold temperature values. Conversely, when molding PMMA at low mold temperature, the higher melt viscosity leads to lower replication. In this condition, water droplets have similar behavior as when they were deposited onto a smooth surface. The surface was not yet practically functionalized. As the mold temperature increases, the replication grade increases, and the surface becomes more hydrophobic. The further increase of mold temperature beneath the T_g of the polymers leads to an increase in replication accuracy that is not followed by the wetting properties of the surface. This can be linked to Cassie–Baxter behavior of the water droplet, which is no longer affected by the depth of the structures on the surface. Indeed, in the Cassie–Baxter state the droplet does not follow the surface topography entirely, but the droplet achieves contact only with the top of the asperities.

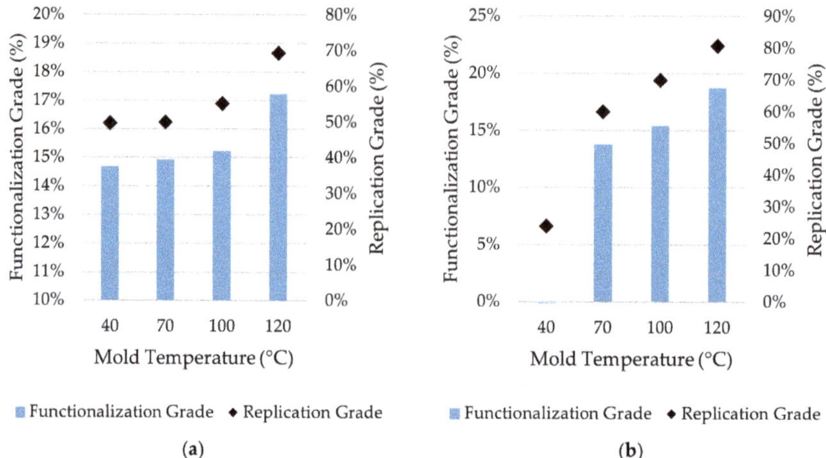

(a)   (b)

Figure 10. Comparison between polymer surface replication grade and functionalization grade for (**a**) PS parts and (**b**) PMMA parts.

4. Conclusions

In this work, different textures were generated by laser ablation and replicated by injection molding with the goal of functionalizing the surface of plastic parts. The process consisted of the fast fabrication of surface structures on steel using a femtosecond laser ablation and their reproduction on plastic parts by µIM. A mold setup with interchangeable inserts was designed, and two different polymers were tested. Mold inserts were treated with a polarized ultrashort pulsed laser source to obtain highly aligned and uniform LIPSSs. The achieved functionalization of the textured surface was studied as a function of the mold temperature. The molded parts were characterized by measuring the degree of replication and their wetting properties.

The textured mold surfaces confirm the suitability of laser technology for scaling up surface functionalization, and every insert was textured in less than 2 min. Tilting the beam angle on the surface is an effective way to manage the morphology of the sub-micron texture, even inducing the formation of a two-period pattern. A loss on the regularity of the pattern occurred when irradiating the surface with higher angles of beam incidence. Further work would be required for the optimization of laser scanning with a tilted beam.

The results of the µIM experimental tests show that the presence of a pattern increases surface functionality. Maximum water contact angle increases of 20% and 17% were found for PMMA and PS samples, respectively. The results show that the achieved functionalization was affected by the polymer temperature, depending on melt viscosity. When molding polymers with a high viscosity temperature dependence, the hydrophobicity of the surface changes rapidly above a threshold value. This temperature was identified between 60 and 70 °C for PMMA.

Further increases of mold temperature did not affect the hydrophobicity, even if the replication accuracy was higher. This phenomenon can be linked to the Cassie–Baxter behavior of the water droplets, which were no longer affected by the depth of the structures. The results for both polymers is attractive from the perspective of producing functionalized low-cost parts. Indeed, keeping a lower mold temperature leads to a decrease in manufacturing costs both from energy consumption and cycle time savings.

Author Contributions: Conceptualization, D.M., M.S.; methodology, D.M., L.P., M.S. and A.B.; laser texturing A.B. and S.D.; injection molding experiments and characterization, M.S. and L.P.; analysis of experimental data, M.S., L.P. and D.M.; supervision, D.M. and M.S.; writing—Original draft preparation, L.P.; writing—Review and editing, D.M., M.S., A.B., G.L. and S.D. All authors have read and agreed to the published version of the manuscript.

Funding: This work was supported by the University of Massachusetts Lowell (Provost Office Start-Up funds to Prof. Davide Masato).

Acknowledgments: Authors acknowledge the contribution of the Core Research Facilities at the University of Massachusetts Lowell for the help with AFM and SEM characterization.

Conflicts of Interest: The authors declare that there is no conflict of interest regarding the publication of this paper.

References

1. Toosi, S.F.; Moradi, S.; Hatzikiriakos, S. Fabrication of Micro/Nano Patterns on Polymeric Substrates Using Laser Ablation Methods to Control Wettability Behaviour: A Critical Review. *Rev. Adhes. Adhes.* **2017**, *5*, 55–78. [CrossRef]

2. Bekesi, J.; Kaakkunen, J.; Michaeli, W.; Klaiber, F.; Schoengart, M.; Ihlemann, J.; Simon, P. Fast fabrication of super-hydrophobic surfaces on polypropylene by replication of short-pulse laser structured molds. *Appl. Phys. A* **2010**, *99*, 691–695. [CrossRef]

3. Xu, Q.; Zhang, W.; Dong, C. Biomimetic self-cleaning surfaces: Synthesis, mechanism and applications. *J. R. Soc. Interface* **2016**, *13*, 20160300. [CrossRef]

4. Sorgato, M.; Zanini, F.; Masato, D.; Lucchetta, G. Submicron laser-textured vents for self-cleaning injection molds. *J. Appl. Polym. Sci.* **2020**, 1–9. [CrossRef]

5. Schulte, V.A.; Diez, M.; Möller, M.; Lensen, M.C. Topography-induced cell adhesion to acr-sP(EO-stat-PO) hydrogels: The role of protein adsorption. *Macromol. Biosci.* **2011**, *11*, 1378–1386. [CrossRef]

6. Shiu, J.Y.; Chen, P. Addressable protein patterning via switchable superhydrophobic microarrays. *Adv. Funct. Mater.* **2007**, *17*, 2680–2686. [CrossRef]

7. Rizzello, L.; Sorce, B.; Sabella, S.; Vecchio, G.; Galeone, A.; Brunetti, V.; Cingolani, R.; Pompa, P.P. Impact of Nanoscale Topography on Genomics and Proteomics of Adherent Bacteria. *ACS Nano* **2011**, *5*, 1865–1876. [CrossRef]

8. Lucchetta, G. Effect of injection molded micro-structured polystyrene surfaces on proliferation of MC3T3-E1 cells. *Express Polym. Lett.* **2015**, *9*, 354–361. [CrossRef]

9. Lourenço, B.N.; Marchioli, G.; Song, W.; Reis, R.L.; Van Blitterswijk, C.; Karperien, M.; Van Apeldoorn, A.; Mano, J.F. Wettability Influences Cell Behavior on Superhydrophobic Surfaces with Different Topographies. *Biointerphases* **2012**, *7*, 1–11. [CrossRef]

10. Davida, C.; Haberlinga, P.; Schnieperb, M.; Sochtigb, J.; Zschokkeb, C. Nano-structured anti-reflective surfaces replicated by hot embossing. *Microelectron. Eng.* **2002**, *61–62*, 435–440. [CrossRef]

11. Zhang, P.; Lv, F.Y. A review of the recent advances in superhydrophobic surfaces and the emerging energy-related applications. *Energy* **2015**, *82*, 1068–1087. [CrossRef]

12. Oberli, L.; Caruso, D.; Hall, C.; Fabretto, M.; Murphy, P.J.; Evans, D. Condensation and freezing of droplets on superhydrophobic surfaces. *Adv. Colloid Interface Sci.* **2014**, *210*, 47–57. [CrossRef]

13. Abdel-Aal, H.A. Functional surfaces for tribological applications: Inspiration and design. *Surf. Topogr. Metrol. Prop.* **2016**, *4*, 1–37. [CrossRef]

14. Gachot, C.; Rosenkranz, A.; Hsu, S.M.; Costa, H.L. A critical assessment of surface texturing for friction and wear improvement. *Wear* **2017**, *372*, 21–41. [CrossRef]

15. Sorgato, M.; Masato, D.; Lucchetta, G.; Orazi, L. Effect of different laser-induced periodic surface structures on polymer slip in PET injection moulding. *CIRP Ann.* **2018**, *67*, 575–578. [CrossRef]

16. Ganesh, V.A.; Raut, H.K.; Nair, A.S.; Ramakrishna, S. A review on self-cleaning coatings. *J. Mater. Chem.* **2011**, *21*, 16304–16322. [CrossRef]

17. Liu, W.; Cai, M.; Luo, X.; Chen, C.; Pan, R.; Zhang, H.; Zhong, M. Wettability transition modes of aluminum surfaces with various micro/nanostructures produced by a femtosecond laser. *J. Laser Appl.* **2019**, *31*, 022503. [CrossRef]

18. Wenzel, R.N. Resistance of solid surfaces to wetting by water. *Ind. Eng. Chem.* **1936**, *28*, 988–994. [CrossRef]

19. Cassie, B.D.; Cassie, A.B.D.; Baxter, S. Of porous surfaces. *Trans. Faraday Soc.* **1944**, *40*, 546–551. [CrossRef]

20. Volpe, A.; Ancona, A.; Trotta, G.; Vázquez, R.M.; Fassi, I.; Osellame, R. Fabrication and assembling of a microfluidic optical stretcher polymeric chip combining femtosecond laser and micro injection molding technologies. *Laser-Based Micro Nanoprocess. XI* **2017**, *10092*, 100920. [CrossRef]

21. Wang, B.; Wang, X.C.; Zheng, H.Y.; Lam, Y.C. Surface modification of polystyrene by femtosecond laser irradiation. *J. Laser Micro Nanoeng.* **2016**, *11*, 253–256.

22. Patel, D.; Jain, V.K.; Ramkumar, J. Micro texturing on metallic surfaces: State of the art. *Proc. Inst. Mech. Eng. Part B J. Eng. Manuf.* **2018**, *232*, 941–964. [CrossRef]

23. Santos, A.; Deen, M.J.; Marsal, L.F. Low-cost fabrication technologies for nanostructures: State-of-the-art and potential. *Nanotechnology* **2015**, *26*. [CrossRef]

24. Orazi, L.; Sorgato, M.; Piccolo, L.; Masato, D.; Lucchetta, G. Generation and characterization of Laser Induced Periodic Surface Structures on plastic injection molds. *Lasers Manuf. Mater. Process.* **2020**, in publication. [CrossRef]

25. Rank, A.; Lang, V.; Lasagni, A.F. High-Speed Roll-to-Roll Hot Embossing of Micrometer and Sub Micrometer Structures Using Seamless Direct Laser Interference Patterning Treated Sleeves. *Adv. Eng. Mater.* **2017**, *19*, 1–8. [CrossRef]

26. Nastulyavichus, A.A.; Kudryashov, S.I.; Saraeva, I.N.; Smirnov, N.A. Nanostructured steel for antibacterial applications. *Laser Phys. Lett.* **2020**, *17*. [CrossRef]

27. Masato, D.; Sorgato, M.; Batal, A.; Dimov, S.; Lucchetta, G. Thin-wall injection molding of polypropylene using molds with different laser-induced periodic surface structures. *Polym. Eng. Sci.* **2019**, *59*, 1889–1896. [CrossRef]

28. Gnilitskyi, I.; Derrien, T.J.Y.; Levy, Y.; Bulgakova, N.M.; Mocek, T.; Orazi, L. High-speed manufacturing of highly regular femtosecond laser-induced periodic surface structures: Physical origin of regularity. *Sci. Rep.* **2017**, *7*, 1–11. [CrossRef]

29. Gurevich, E.L. Mechanisms of femtosecond LIPSS formation induced by periodic surface temperature modulation. *Appl. Surf. Sci.* **2016**, *374*, 56–60. [CrossRef]

30. Bonse, J.; Höhm, S.; Kirner, S.; Rosenfeld, A.; Krüger, J. Laser-induced Periodic Surface Structures (LIPSS)—A Scientific Evergreen. *Conf. Lasers Electro-Opt.* **2016**, *23*, STh1Q.3. [CrossRef]

31. Prokhorov, A.M.; Svakhin, A.S.; Sychugov, V.A.; Tishchenko, A.V.; Khakimov, A.A. Excitation and resonant transformation of a surface electromagnetic wave during irradiation of a solid by high-power laser radiation. *Sov. J. Quantum Electron.* **1983**, *13*, 568–571. [CrossRef]

32. Hansen, H.N.; Hocken, R.J.; Tosello, G. Replication of micro and nano surface geometries. *CIRP Ann. Manuf. Technol.* **2011**, *60*, 695–714. [CrossRef]

33. Maghsoudi, K.; Jafari, R.; Momen, G.; Farzaneh, M. Micro-nanostructured polymer surfaces using injection molding: A review. *Mater. Today Commun.* **2017**, *13*, 126–143. [CrossRef]

34. Masato, D.; Sorgato, M.; Lucchetta, G. Analysis of the influence of part thickness on the replication of micro-structured surfaces by injection molding. *Mater. Des.* **2016**, *95*, 219–224. [CrossRef]

35. Lin, H.Y.; Chang, C.H.; Young, W.B. Experimental and analytical study on filling of nano structures in micro injection molding. *Int. Commun. Heat Mass Transf.* **2010**, *37*, 1477–1486. [CrossRef]

36. Lucchetta, G.; Sorgato, M.; Masato, D. Vacuum-Assisted Micro Injection Molding. In *Micro Injection Molding*; Tosello, G., Ed.; Hanser Publications: Munich, Germany, 2018.

37. Liparoti, S.; Speranza, V.; Pantani, R. Replication of Micro- and Nanofeatures in Injection Molding of Two PLA Grades with Rapid Surface-Temperature Modulation. *Materials* **2018**, *11*, 1442. [CrossRef]

38. Sorgato, M.; Masato, D.; Lucchetta, G. Effect of vacuum venting and mold wettability on the replication of micro-structured surfaces. *Microsyst. Technol.* **2017**, *23*, 2543–2552. [CrossRef]

39. Batal, A.; Michalek, A.; Garcia-Giron, A.; Nasrollahi, V.; Penchev, P.; Sammons, R.; Dimov, S. Effects of laser processing conditions on wettability and proliferation of Saos-2 cells on CoCrMo alloy surfaces. *Adv. Opt. Technol.* **2019**, 8062. [CrossRef]

40. Wang, X.; Duan, J.; Jiang, M.; Ke, S.; Wu, B.; Zeng, X. Study of laser precision ablating texture patterns on large-scale freeform surface. *Int. J. Adv. Manuf. Technol.* **2017**, *92*, 4571–4581. [CrossRef]

41. *ISO 25178-2 Geometrical Product Specifications (GPS)—Surface Texture: Areal—Part 2*; International Organization for Standardization: Geneva, Switzerland, 2012.

42. Razi, S.; Ghasemi, F. Laser-assisted generation of periodic structures on a steel surface: A method for increasing microhardness. *Eur. Phys. J. Plus* **2018**, *133*, 49. [CrossRef]
43. Liou, A.C.; Chen, R.H. Injection molding of polymer micro- and sub-micron structures with high-aspect ratios. *Int. J. Adv. Manuf. Technol.* **2006**, *28*, 1097–1103. [CrossRef]

 *micromachines*

Article

Characterization of Stereolithography Printed Soft Tooling for Micro Injection Molding

Daniel Dempsey, Sean McDonald, Davide Masato  **and Carol Barry ***

Department of Plastics Engineering, University of Massachusetts Lowell, Lowell, MA 01854, USA;
Daniel.dempsey@newbalance.com (D.D.); Sean_McDonald@student.uml.edu (S.M.);
davide_masato@uml.edu (D.M.)
* Correspondence: Carol_Barry@uml.edu; Tel.: +1-(978)-934-3436

Received: 29 June 2020; Accepted: 27 August 2020; Published: 28 August 2020

Abstract: The use of microfeature-enabled devices, such as microfluidic platforms and anti-fouling surfaces, has grown in both potential and application in recent years. Injection molding is an attractive method of manufacturing these devices due to its excellent process throughput and commodity-priced raw materials. Still, the manufacture of micro-structured tooling remains a slow and expensive endeavor. This work investigated the feasibility of utilizing additive manufacturing, specifically a Digital Light Processing (DLP)-based inverted stereolithography process, to produce thermoset polymer-based tooling for micro injection molding. Inserts were created with an array of 100-μm wide micro-features, having different heights and thus aspect ratios. These inserts were molded with high flow polypropylene to investigate print process resolution capabilities, channel replication abilities, and insert wear and longevity. Samples were characterized using contact profilometry as well as optical and scanning electron microscopies. Overall, the inserts exhibited a maximum lifetime of 78 molding cycles and failed by cracking of the entire insert. Damage was observed for the higher aspect ratio features but not the lower aspect ratio features. The effect of the tool material on mold temperature distribution was modeled to analyze the impact of processing and mold design.

Keywords: micro injection molding; additive manufacturing; stereolithography; replication

1. Introduction

Micro-structured surfaces hold the potential to improve the inherent functionality of products ranging from consumer goods to medical devices to adhesive technology. Whether these systems work based on surface energy modification, light wave manipulation, or microfluidic transport, the inclusion of periodic or asymmetric positive and negative microscale features onto the surfaces of products can yield new properties which would not be intrinsic to an unstructured product. Many researchers have worked at integrating these types of technologies into applications that will have daily use, with a significant focus on surfaces which are inherently super-hydrophobic [1], antireflective [2], and antimicrobial [3]. Researchers also are developing new classes of products, with an example being microfluidic platforms. In general, actual products utilizing surface engineering techniques are somewhat limited, and specifically, the commercialization of microfluidic devices has remained challenging. Issues have been a lack of standards for device testing and development [4], general market dynamics being unfavorable for investment [5], and practices which use processes and techniques for device creation which, at the laboratory scale, are simple and straightforward operations yet ultimately are not feasible to scale for mass manufacturing methods [6].

Table 1 lists techniques and processes which are commonly utilized to produce microfluidic platforms, both on the laboratory scale and for possible commercial mass manufacturing. Devices have been machined directly using micromachining and laser machining techniques, but the minimum feature sizes typically are of around 100 µm [7,8]. While casting of polydimethylsiloxane (PDMS)—i.e., soft lithography—requires low temperatures and pressures, the workflow utilizes labor-intensive techniques, making the process more suitable for laboratory scale development of devices. Hot embossing, roller imprinting, and injection molding provide higher throughputs but require higher temperatures (T) and pressures (P), as well as greater capital investment. Additionally, these higher throughput workflows often require a series of expensive and time-consuming techniques for the creation of the tooling with micro-structured surfaces [6]. If the development bottlenecks in the higher throughput processes could be solved, such that more iteration could be done on the laboratory scale, this would be an excellent path forward for eventual ease of commercialization.

Table 1. Comparison of manufacturing processes for micro-featured device manufacturing.

Process	Resolution (µm)	Cycle (min)	Throughput (pcs/mo.)	T (°C)	P (MPa)	Ref.
CNC machining	100	5–30	300–2000	–	–	[7]
Laser machining	100	5–30	300–2000	–	–	[7]
PDMS casting	<1	30–60	150–300	100	0.1	[9]
Hot embossing	<1	10–30	300–1000	100–250	5	[9]
Roller imprinting	<1	0.15–0.5	5000+	100	1	[10]
Injection molding	<1	0.15–0.5	5000+	100–250	10–30+	[7]

In injection molding of microfluidic devices, a tooling insert with the microchannel layout is commonly employed in the cavity to form the features on the part surface [11–13]. After the predetermined amount of polymer is injected into the mold, pack pressure applied to the melt typically completes the formation of the features. To attain higher replication of microfeatures, the process usually requires elevated melt temperatures [13–15] and high injection pressures and velocities [16] to decrease the viscosity of the polymer through shear heating and to delay the solidification of the melt as it comes into contact with the mold wall [15]. Additionally, many researchers believe that mold temperature is the highest temperature setting effect on the process and recommend the mold temperature be higher than the glass transition temperature of the polymer being injected [14,16]. The high temperatures and pressures, however, place considerable stress on the tooling inserts.

When considering the manufacturing processes utilized for the production of microfeature-containing injection molding inserts, a change in design or design complexity is a complicated and expensive endeavor. Over the last 20+ years, many researchers have investigated a variety of materials to employ for tooling inserts in the microinjection molding process, utilizing a wide array of manufacturing techniques [17]. Among the most prevalent in research are silicon produced with x-ray, e-beam, and UV lithography [18], SU-8 epoxy formed via UV lithography [19], hybrid tooling approaches of imprinted polyimide with metalized surfaces [20], electroformed and chemically-etched nickel [21], micro-milled [22], and chemically-etched stainless steel [23], and micro-milled and cast bulk metallic glass [6]. Each combination of tooling material and manufacturing process yields varying capabilities for minimum feature resolutions, lifecycle capabilities, and required lead times for tooling production. Table 2 shows a comparison of these production methods with a view to production scalability.

Current tooling methods fall into one of two categories: (1) fast to produce but limited cycle capability and (2) long lead time for production but high cycle capacity. Further exacerbating the issue is the production method, where each long lead time tool is produced via a process that traditionally makes a single part at a time. Accordingly, scaling a process becomes difficult as parallelization of production requires a high number of tooling inserts. Moreover, for all production methods, a change in design or a desire to evaluate multiple designs becomes problematic as either the window to evaluate is small (low cycle tools) or the lead time to evaluate is large (high cycle tools). Zhang et al. concluded that the more robust materials, such as stainless steel, nickel, and bulk metallic glass, were preferable for molding but encountered challenges from a tool manufacturing standpoint [6]. Resolution limits prevent the use of stainless steel [12,23], void formation in electroformed nickel limits cycles [24], and overall mechanical properties and cost could prevent full adoption of bulk metallic glass [6]. Therefore, the capability to quickly produce tooling inserts with sufficient mechanical properties to withstand the elevated temperatures and pressures of micro injection molding becomes a critical factor to efficiently prototype and scale products.

Table 2. Comparison of technologies for the manufacturing of micro-structured injection molds (adapted from [6]).

Tool Material	Common Manufacturing Method	Minimum Resolution (µm)	Manufacturing Time	Tool Life (Cycles)
SU8	UV Lithography	25	0.5 h	Low (<100)
Silicon	UV Lithography	25	4.5 h	Low (<100)
Hybrid Tooling	Embossing	25	2 days	20–50
Stainless Steel	Micro Milling, Chemical Etching	25	5 days	50,000+
Nickel	Electroforming	5	1–2 weeks	Low (1000)
Bulk Metallic Glass	Micro Milling, Casting	25	5 days	20,000

Additive manufacturing offers the opportunity to merge ease of design changes with an agile manufacturing process. Table 3 outlines the various additive manufacturing technologies broken down by substrate material and process type. When investigating the additive production for injection molding inserts, intuitively, one would gravitate towards metal printing methods. Indeed, there is prior art for researchers utilizing metal printing commonly for the production of inserts with conformal cooling channels [25,26]. At the microscale, however, commercial metal printing processes prove challenging. Generally, metal processes are inhibited by two factors: (1) limited capable resolution due to minimum spot diameters of lasers [27] and (2) high surface roughness due to powder particle size [28].

When Vasco and Pouzada [29] utilized electron beam melting (EBM) to print an injection molding insert containing positive and negative micro walls (200 µm high by 100 µm wide) arrayed in pentagon and star patterns, they noted incomplete feature printing due to the minimum spot size of the laser. This limitation is shared by all of the laser based metal printing processes as they exhibit limited laser focus diameters of 50–300 µm [27]. After molding with PP, Vasco and Pouzada [29] also reported prevalent stair-stepping on part sidewalls due to printing stratification. Furthermore, Mendible et al. [28] evaluated the direct metal laser sintering (DMLS) process to print bronze casting mold inserts for injection molding and observed a high degree of surface roughness (40–60 µm) when compared to a traditionally machined version. Nagahanumaiah et al. [30] showed that polishing a DMLS bronze tool (for a macroscale gear hub) would allow for thousands of molding cycles in nylon with no tool wear; however, polishing techniques would need to be evaluated for positive tooling microfeatures because the gear hub had a contoured surface with no projections.

While commercial metal printing processes are certainly limited by total resolution for micro-manufacturing applications, polymer processes are split into two groups: (1) print methods which share resolution limits with metal counterparts and (2) processes with acceptable resolution but insufficient material properties. Selective laser sintering (SLS) and multi-jet fusion (MJF) are two examples where resolution limitations are similar to EBM and DMLS; surface roughness can be challenging as typical particle sizes range from 15 to 150 µm [31]. Resolution challenges also hamper fused deposition modeling (FDM), where the print stratification effects and minimum resolution are both a function of the filament diameter deposited by the printer, often limited to a minimum resolution of 100 µm [32]. Furthermore, the filament deposition methods leave voids between perimeter and infill passes, which are not conducive to the necessary compression strength required for injection mold tooling.

For systems which do exhibit acceptable process resolution, material properties often are the challenge. Material jetting (PolyJet) would be an attractive technology, with resolution limits around 80 μm and smooth surface finishes [33]; however, when utilized in studies to create short-run injection tooling, the material heat deflection temperature (HDT) becomes a liability, with all materials exhibiting HDTs under 95 °C [34]. For a PolyJet insert, Mendible et al. [28] reported drastically increased cycle times compared to a traditional machined insert—i.e., 200 and 45 s, respectively—dimensional instability in molded parts within 10 cycles, and tooling life of 130 shots before complete failure. VAT photopolymerization processes show promise in terms of resolution with laser-based SLA systems and DLP-based SLA systems, reporting minimum features of 76 [34] and 19 μm [35], respectively. The differences in X-Y resolution in SLA are due to the choice of print engines: (1) laser-based resolution is defined by the laser spot size, and (2) DLP-based resolution is a function of the pixel size of the projector. For SLA, material development is more advanced than with filament or powder-based methodologies, and some materials have higher heat deflection temperatures (HDT) and flexural modulus properties than PolyJet formulations. When Gheisari et al. [36] employed a DLP-based SLA system to create three tooling inserts for microcantilevers with thicknesses of 30, 120, and 275 μm, they observed failure across three inserts all within five injection molding cycles with polypropylene. They also did note an incorrect print orientation, which may have led to over cure and thus a deterioration in mechanical properties. Vasco and Pouzada [29] also produced star and pentagon featured tooling through a laser-based SLA system, yielding a spot size of 75 μm; initial results were affected by early breakage of microfeatures during the injection cycle, although tool duration was not reported and investigation of wear properties for the SLA tooling was abandoned. Their channel designs had a higher high aspect ratio (AR = 2:1), which could have contributed to stress on the tooling.

Table 3. Commercial additive manufacturing processes and resolution limits.

Substrate Material	Additive Manufacturing Process	Minimum Resolution (μm)	Refs.
Polymer Processes	Stereolithography	75	[34,35]
	Digital Light Processing	15	[34,35]
	Continuous Digital Light Processing	75	[34,35]
	Fused Deposition Modeling	100	[32]
	Material Jetting	80	[33]
	Multi Jet Fusion	50–150	
Metal Processes	Selective Laser Sintering	50–300	[27,28]
	Direct Metal Laser Sintering	50–300	
	Electron Beam Melting	50–300	
	Laser Engineering Net Shape	50–300	
	Electron Beam Additive Manufacturing	50–300	

The objective of this work was to investigate the feasibility of creating a lower aspect ratio microfeature-enabled tooling for the injection molding process using commercially available additive manufacturing technologies. A DLP-based inverted stereolithography system was utilized to print tooling inserts with an array of 100-μm wide positive microchannels with heights of 24, 47, 71, and 94 μm (creating respective aspect ratios of 0.7, 0.59, 0.39, 0.2) on the surface. Simple microfluidic channels were replicated in high flow polypropylene, and tooling wear was evaluated using a combination of contact profilometry, optical microscopy, and scanning electron microscopy. Tooling life was investigated by molding samples until the tooling fractured during injection molding.

2. Materials and Methods

2.1. Microstructured Tooling Design

The micro cavities in the injection molding tooling insert were designed to mimic the features that characterize a microfluidic chip. Figure 1 shows the design of one of the tooling inserts and the location of the microfeatures on its top surface. Different microstructures were designed with different sizes and aspect ratios. Each group of features had a nominal channel width and pitch of 100 μm. However, the structures in the tooling inserts had a height of (A) 25 μm, (B) 50 μm, (C) 75 μm, and (D) 100 μm. Indeed, variation in features height allowed evaluation of the effect of aspect ratio on replication, where researchers identified this as the most significant parameter for replication of microstructured surfaces by injection molding [37]. The tooling inserts had a 14 mm square overall size, with thickness of 10 mm.

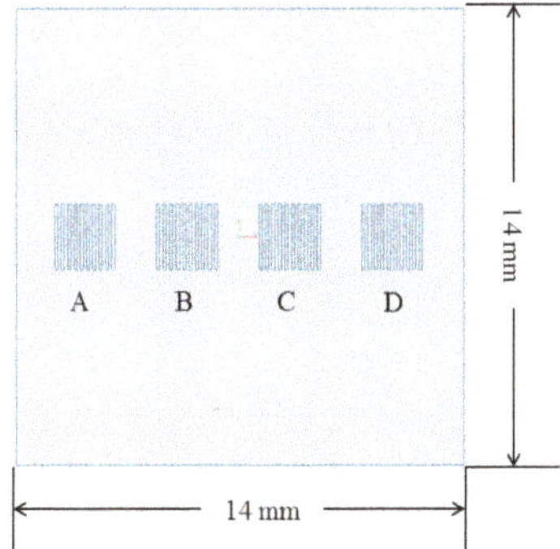

Figure 1. Design of the tooling insert and positioning of the microfeatures.

2.2. 3D Printing of Tooling Inserts

The tooling inserts were produced via inverted stereolithography on a commercially available system (Perfactory 4 Mini, EnvisionTEC GMBH, Gladbeck, Germany). The machine has a maximum build size of 38 × 24 × 220 mm, pixel size of 19 μm, and a z-axis resolution of 15–150 μm. In this work, an inverted SLA was selected over traditional SLA because it allows (i) use of resins with improved mechanical properties and (ii) cheaper production with smaller vat requirements [38].

The print time for all tooling inserts was 4 h and 12 min, and each insert was UV flood post-cured for 4.5 min. Figure 2 shows one of the printed tooling inserts and details of the microfeatures. A total of 10 inserts were produced for this study; all inserts were printed simultaneously in a single build. This parallelization of tooling production is considered an advantage of the DLP-SLA system.

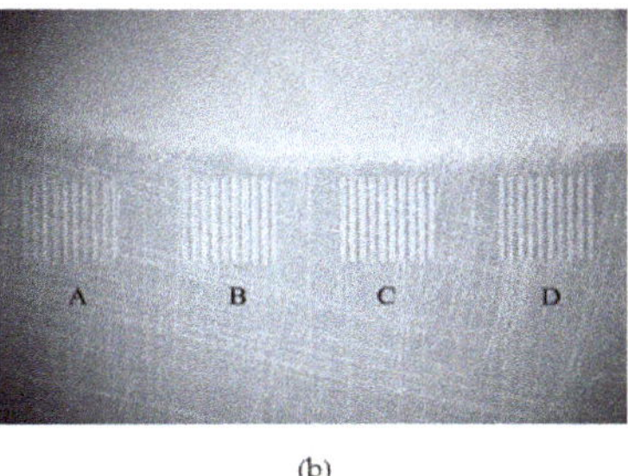

(a)   (b)

Figure 2. Photos of stereolithography (SLA) 3D printed tooling inserts: (**a**) bulk of the insert and surface microfeatures and (**b**) top view of the printed microfeatures.

The resin used to print the inserts was a cross-linkable acrylic-epoxy formulation (HTM 140 V3), whose mechanical properties are reported in Table 4. This formulation was chosen because of its high heat deflection temperature and flexural modulus and the ability to print at a layer thickness of 15 μm. The mechanical properties of HTM 140 V3 suggested its suitability for withstanding the higher pressures and temperatures in the micro injection molding environment.

Table 4. Main properties of the resin selected to print the micro molding tooling inserts [28].

Property	Units	Value
Tensile Strength	MPa	56
Elongation at Break	%	3.5
Flexural Strength	MPa	115
Flexural Modulus	MPa	3350
Heat Deflection Temperature	°C	140

2.3. Injection Molding Setup

The tooling inserts printed using SLA were mounted into a mold base using a steel carrier insert. A support pillar and two sheets of 0.4-mm thick polytetrafluoroethylene (PTFE) were used to prevent deflection of the 3D printed tooling insert due to high pressure from the melt polymer. Figure 3a shows an exploded view of the mold base and assembly of the 3D printed tooling insert, which was mounted into the B-plate with the features facing the A-side of the mold. The rest of the mold geometry was machined into the steel B-plate of the mold. The mold was used to manufacture a plastic part with the geometry shown in Figure 3b. The part is a center-gated disk with a thickness of 2 mm and a diameter of 25 mm. The microstructured area was offset to one side of the sprue to provide for parallel flow across the microfeatures. The polymer utilized was a high flow polypropylene (Pinnacle Polymers, Pinnacle PP 1345Z, Garyville, LA, USA), which has a melt flow rate (MFR) of 45 g/10 min.

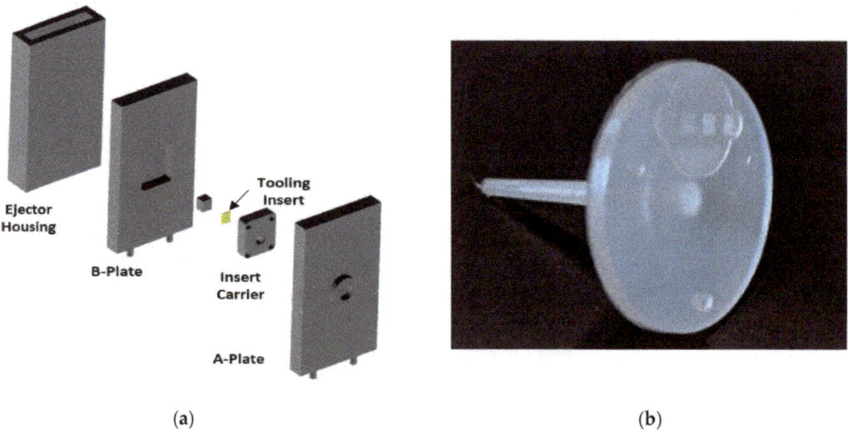

Figure 3. (a) Exploded view of the mold base shows the assembly of the 3D printed tooling insert in the steel mold and (b) plastic part injection molded using the 3D printed tooling insert and replicated microfeatures.

Injection molding trials were carried out using a 3-ton Nissei micro injection molding machine (Model: AU3E). The machine is characterized by a two-stage injection unit with a 14-mm diameter metering screw and an 8-mm diameter injection plunger. Table 5 reports the main injection molding processing conditions, which were kept constant for all experiments. In particular, temperatures were selected according to indications from resin manufacturer; injection velocity and velocity switchover point were optimized through previous work [12]; packing conditions and cooling conditions were adjusted to eliminate sink mark formation.

Table 5. Process molding conditions adopted for the experimental campaign.

Parameter	Unit	Value
Mold Temperature	°C	65
Barrel Temperature	°C	195
Shot Size	mm	12.5
Injection Velocity	mm/s	80
Pack Pressure	MPa	20
Pack Time	s	15
V/P Switchover Point	MPa	15
Cooling Time	s	10

The main goal of the injection molding trials was the evaluation of tool life for the SLA 3D printed tooling inserts. Table 6 reports the plan for the injection molding experiments that were carried out using the 10 SLA printed tooling inserts. The evaluation cadence was broken up to yield a view of before and after molding insert conditions and in-process part replication performance across an increasing number of cycles.

Table 6. Injection molding experimental plan for 3D printed tooling inserts life investigation.

Tool	Cycle Evaluation	Part Measurement (Cycles)	Tool Measurement (Cycles)
1	Process Optimization	-	-
2	20	20	20
3	40	20, 40	40
4	60	20, 40, 60	60
5	80	20, 40, 60, 80	80
6	100	20, 40, 60, 80, 100	100
7	200	50, 100, 150, 200	200
8	300	75, 150, 225,300	300
9	500	100, 200, 300, 400, 500	500
10	1000	200, 400, 600, 800, 1000	1000

2.4. Characterization of Replicated Topography

Scanning electron microscopy (SEM, JEOL JSM-7401F, Akishima, Tokyo, Japan) was initially utilized for a qualitative evaluation of microchannels replication. For SEM analyses, samples were sputter-coated with a 10-Å thick layer of gold to permit greater electronegativity. SLA printed tooling inserts were also observed before and post injection molding experiments using confocal microscopy (Zeiss, Stemi 2000 CS, Oberkochen, Germany).

The replication accuracy of the injection molding process carried out with the SLA printed tooling inserts was evaluated by metrological characterization of the replicated structures and comparison with their nominal tool dimension. The height of the replicated microstructures was acquired using a contact profilometer (Bruker XT, Billerica, MA, USA). The height of the replicated structures was measured by monitoring the force transmittance of a stylus as it was dragged across the topography of the sample. The difference between the top surface of the plastic part and the bottom of the structures replicated on the tooling insert was defined as the replicated depth. From channel depth measurements, the feature depth ratio (*DR*) was calculated as

$$DR = \frac{d_p}{d_t} \tag{1}$$

where d_p is the average depth of the replicated channel in the plastic part, and d_t is the average measured height of the tooling projection.

3. Modeling

The injection molding process with the SLA mold was modeled using Moldex3D (CoreTech System Co., Hsinchu County, Taiwan) Studio R17 to understand the effect of the 3D printed mold material on processing. The numerical simulation solves the following system of governing equations to determine the characteristics of the polymer flow [39]. The system includes the conservation of mass (Equation (2)), of the linear momentum (Equation (3)), and of energy (Equation (4)):

$$\frac{d}{dt}\rho + \nabla \cdot (\rho v) = 0 \tag{2}$$

$$\rho \frac{d}{dt} v = \rho g - \nabla P + \eta \nabla^2 v \tag{3}$$

$$\rho c_p \left(\frac{\partial}{\partial t} T + v \nabla T \right) = \beta T \left(\frac{\partial}{\partial t} P + \vec{v} \times \vec{\nabla} P \right) + \nabla \cdot (k \nabla T) + \eta \dot{\gamma}^2 \tag{4}$$

where ρ is the density, η the melt viscosity, c_p is the specific heat, β is the heat expansion coefficient, k is the thermal conductivity, t is the time, v is the velocity vector, g is the gravitational acceleration constant, P is the hydrostatic pressure, T is the temperature, and $\dot{\gamma}$ is the shear rate. Moldex3D solves the governing equations using the high performance finite volume method (HPFVM), which is used to subdivide the computational domain into a finite number of non-overlapping control volumes. For each control volume, the transport variables are calculated at the center of the element; then, the upwind scheme is used to approximate the variables at the faces [40].

The aim of injection molding simulation was that of understanding temperature distribution during the filling of the microstructures. Indeed, when using a plastic 3D printed mold, it is crucial to understand the effect of the lower mold thermal conductivity on the injection molding process. The injection molding simulations were run designing different models with different mold material properties. In particular, the 3D printed tool was compared to a steel and an aluminum tool. Table 7 lists the main properties of the mold materials that were used for the analysis.

Table 7. Main properties of the different mold materials used for modeling.

Property	Unit	3DP Plastic	Aluminum 6061	P20 Steel
Density	g/cm3	1.3	2.7	7.8
Heat Capacity	J(KgK)	1030	896	462
Thermal Conductivity	W/(mK)	0.3	154	29
Elastic Modulus	Pa	3.4×10^3	6.9×10^{10}	2.1×10^{11}
Poisson Ratio	-	0.35	0.33	0.30
Coefficient of Linear Thermal Expansion	1/K	3.0×10^{-4}	2.2×10^{-5}	1.3×10^{-5}

The CAD model of the part with the microstructures was imported into Moldex3D and meshed using a boundary layer method (BLM) approach (Figure 4a). Seeding was initially carried out to divide the surface of the model with a global edge length of 0.8 mm. Then, the edges of the microstructures were selected to define a smaller element edge size, i.e., 0.025 mm. This allowed having four elements along the thickness of the smaller surface features (Figure 4b). The volume mesh was then created by combining prism elements for the outer layer and tetrahedron in the core of the part. The mold base CAD model was imported and meshed using 3D tetrahedron elements. A mesh size of 2 mm was selected for the mold base, with smaller elements at the interface with the part. The convection between the mold and the environment (i.e., external thermal boundary condition) was defined by specifying an air temperature of 25 °C. The thermal boundary condition at the part/tool interface was automatically calculated by the proprietary software algorithm, which considers the process parameters, material properties, and model geometry.

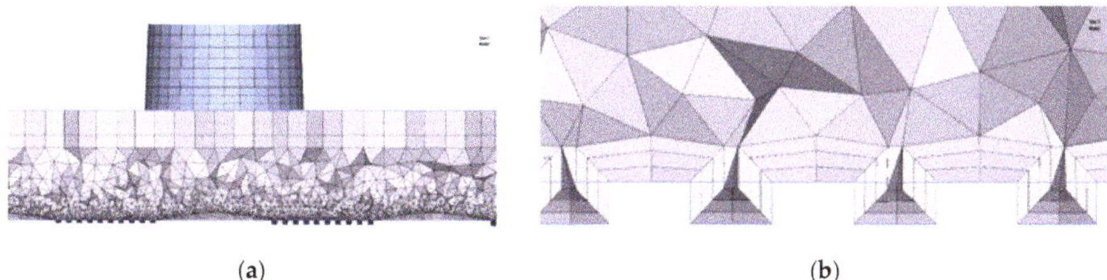

(a)   (b)

Figure 4. Meshing of (**a**) the part and (**b**) details of the elements in the microstructures.

The polypropylene material database was selected from the software library and used to model the melt polymer flow. Processing conditions were set according to experimental information and machine limitations.

4. Experimental Results

4.1. Dimensional Characterization of 3D Printed Tooling Inserts

Table 8 presents the dimensions of the micro-structured tooling inserts printed using the inverted SLA process. The dimensions obtained in the tooling were on average 19% wider and 6% shorter than the nominal design dimensions. Thus, the aspect ratio of the microfeatures in the tooling inserts was on average 20% smaller than the design specifications. This deviation in the printed geometry in inverted SLA printing was attributed to two factors: (1) overexposure and compression of initial layers, which causes shortened Z-direction dimensions and (2) the resolution limits of the system chosen for production. When using an inverted SLA process, the first few layers are overexposed to ensure that there is excellent adhesion to the build plate as the part is printed. This overexposure causes the initial layers to be thinner than the standard layer thickness, which has the effect of offsetting all Z height measurements by the compression amount [41].

Table 8. Characterization of microfeatures' dimensions on 3D printed tooling inserts and comparison with nominal design dimensions.

Group	Nominal Dimensions			Measured Dimensions		
	Width (µm)	Height (µm)	Aspect Ratio	Width (µm)	Height (µm)	Aspect Ratio
A	100	25	0.25	119	23.5	0.20
B	100	50	0.50	119	47.0	0.39
C	100	75	0.75	119	70.5	0.59
D	100	100	1.00	119	94.0	0.79

Regarding the wider-than-designed channels, the increase of 19 µm falls within the resolution of a voxel for the Perfactory Mini system with a 75 mm lens and well within the accuracy of the system. The overgrowth of channels could be eliminated by performing a calibration to the print engine, ensuring perfect alignment with the pixels of the projector, or, at worst, scaling down future geometry to anticipate the XY growth in printing, although pixel growth may still be likely.

4.2. Topography Characterization of 3D Printed Tooling Inserts

As illustrated by the scanning electron microscopy (SEM) images and contact profilometry in Figure 5, the overall quality of the printed channels was excellent, with near full fidelity of the channel depths. The qualitative SEM analysis indicated that the 3D printed microstructures were regular and homogenous across the whole printing area, with the absence of significant surface and morphological defects (Figure 5a,b). It is interesting to note that the contact profilometry trace (Figure 5c) indicates that the tops of the channels exhibited a slight rounding effect. This factor was not unexpected as ERM (EnvisonTEC's Enhanced Resolution Mode), which physically shifts the print engine by one half pixel to essentially deposit two layers in place of one on the part surface, was not used for this work. At the micrometer scale, merely utilizing the printer's anti-aliasing function was enough to create rounded channel tops. Similar to applications in 2D printing, in the DLP system, surface voxels are exposed on a gradient of 255 shades, with 255 being pure white and 0 being black [42]. The variations in exposure energy lead to a smooth surface finish and, in this work, a rounded microchannel. A graphical comparison of traditional layer-wise stratification, which produces steps in the microfeatures and the anti-aliasing effect that created the rounded features is presented in Figure 6. Some rounding also may have been due to limitations of the contact profilometry measurements.

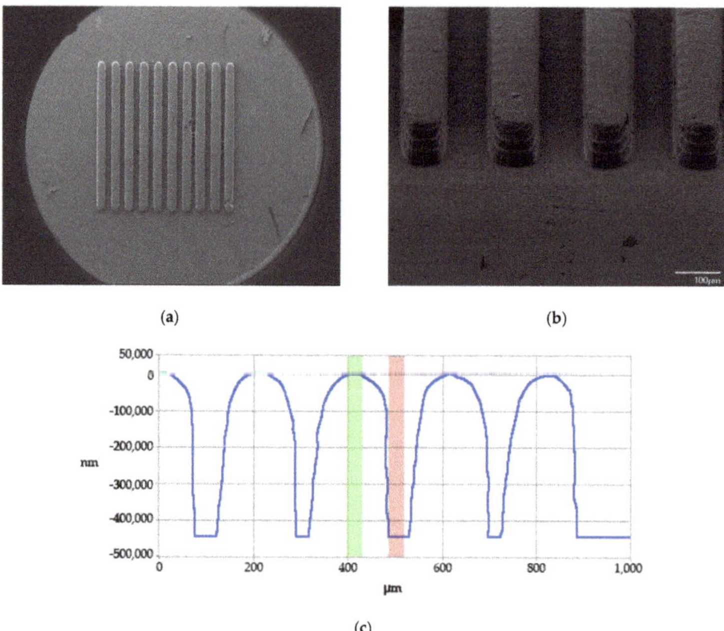

Figure 5. Group B tooling: (a) SEM image of the top of features, (b) SEM image of isometric view of features, and (c) contact profilometry trace of these features.

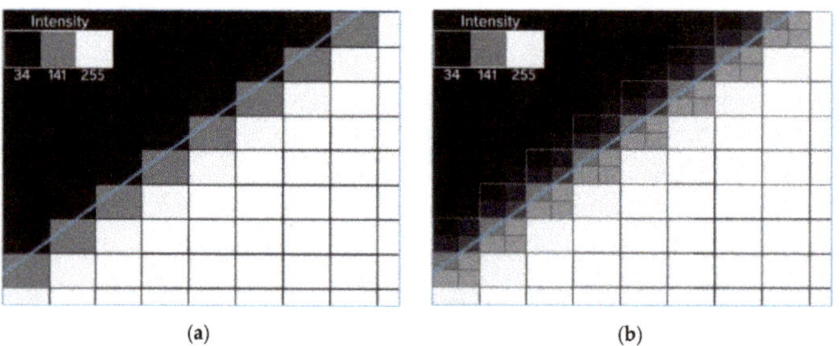

Figure 6. Graphical representation of (**a**) traditional layer-wise stratification and (**b**) DLP anti-aliasing strategy [42].

4.3. Injection Molding Process Optimization

The use of 3D printed tooling inserts with microfeatures required process optimization before the tool life for the different inserts could be characterized. Indeed, when molding parts with microstructured surfaces, semi-crystalline materials like polypropylene exhibit higher and more unpredictable levels of shrinkage than amorphous polymers such as polymethylmethacrylate (PMMA) and polycarbonate (PC) [43,44]. The initial trials were run with a barrel temperature of 215 °C, a packing pressure of 10 MPa, and a pack time of 10 s. As shown in Figure 7a,b, these parts exhibited sink marks on the microfeatured surface and the part surface. These sink marks distorted the molded microfeatures (Figure 7b). Elimination of the sink mark required two rounds of process optimization. First, the melt temperature was lowered from 215 °C to 195 °C to reduce shrinkage of the polypropylene melt; this molding was performed with Tool 1. Second, the pack pressure and pack time were increased from 10 MPa and 10 s to 20 MPa and 15 s, respectively, to allow greater compensation for shrinkage. The sink marks were not present with these trials (which were performed with Tool 2). Parts molded with Tools 3, 4, and 5 employed the optimized injection molding conditions and also exhibited no sink marks. Figure 7c illustrates Groups B and C for a Tool 4 part, confirming that process changes had eliminated the defect.

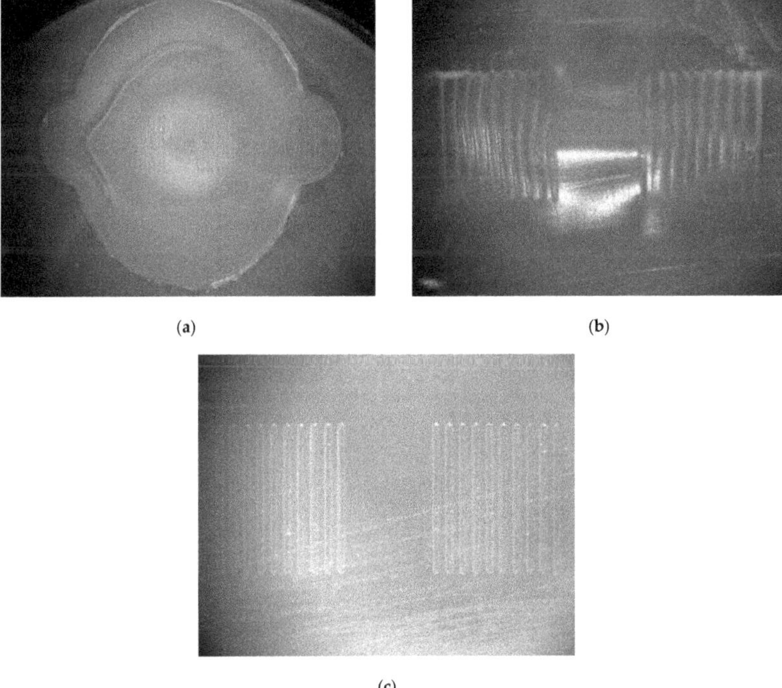

(a)　　　　　　　　　　(b)

(c)

Figure 7. Optical micrographs of (**a**) microfeature molded surface, (**b**) Groups B and C molded with initial processing conditions that produced sink marks, and (**c**) Groups B and C molded with optimized process conditions, which eliminated the sink marks.

4.4. Characterization of 3D Printed Tool Life

The life of the tooling inserts was analyzed using optical imaging of the 3D printed parts at different intervals during the injection molding trials. Different failure behavior was observed, either for the whole insert or for the surface microfeatures. Figure 8 shows images of different failure modes in the 3D printed tooling inserts.

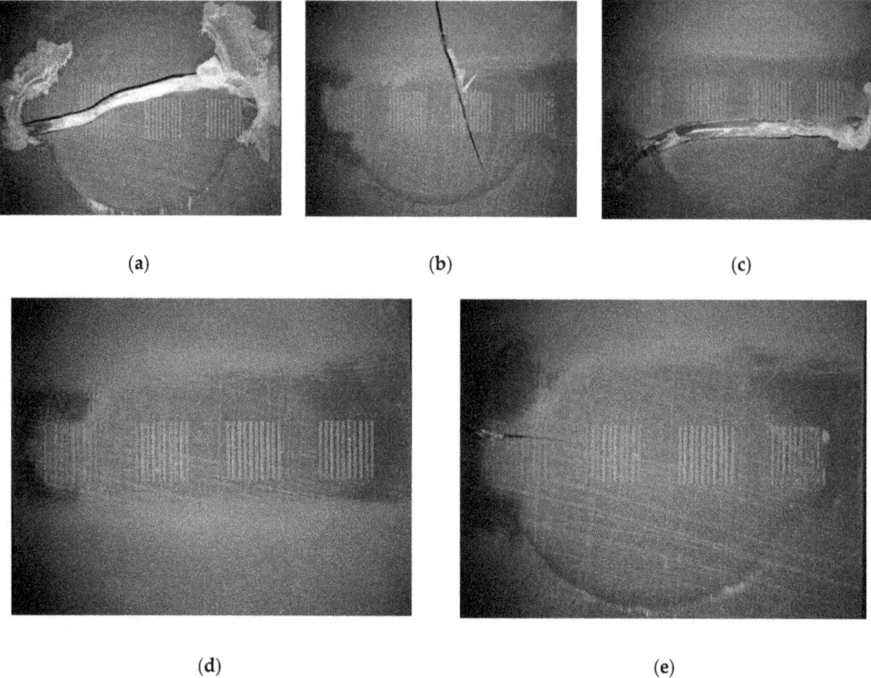

Figure 8. Images of tooling inserts: (a) Tool 1 after 36 shots, (b) Tool 2 after 16 shots, (c) Tool 3 after 29 shots, (d) Tool 4 after 60 shots (not cracked), and (e) Tool 5 after 78 shots.

During the optimization of the process settings, the inserts proved to be more brittle than anticipated. Tool 1 failed at 36 shots, whereas Tools 2 and 3, which were exposed to higher pack pressures for longer times, failed at 16 and 29 shots, respectively. As shown in Figure 9a–c, the microfeatures in these parts generally did not fail, but rather the entire inserts cracked across the center. With optimized processing conditions, Tool 4 lasted 60 shots without failure and was pulled for tool wear evaluation. Tool 5 failed after 78 molding cycles.

Figure 9d presents a close-up view of Group D features, which had the highest aspect ratio (i.e., 0.79). These positive microfeatures on the surface exhibited consistent brittle failure across all trialed samples. In contrast, the microfeatures in Groups A, B, and C, with aspect ratios 0.20, 0.39, and 0.59, respectively, survived the molding process until bulk tooling failure, with no evidence of brittle failure (Figure 9a–c). The results in Figure 9 are also consistent with the brittle failure of higher aspect ratio positive features experienced in prior work with DLP [29] and SLA [36] molds. Overall, the wear of the features suggests that there may be an aspect ratio limit with this tooling system. Further exploration of SLA formulations, with increased flexural modulus, may be required for higher aspect ratio microfeature durability.

The SEM image20 of Group D features from Tool 5 (Figure 10) does indicate some surface deterioration, which can be attributed to thermal wear and failure during ejection. Group D structures failed because of their aspect ratio (i.e., 0.79), which leads to higher and weaker replicated microstructures that are broken during the ejection phase [45]. This is accentuated by the high melt temperature (i.e., 195 °C), which, during the process, is above the HDT for the tooling insert (i.e., 140 °C).

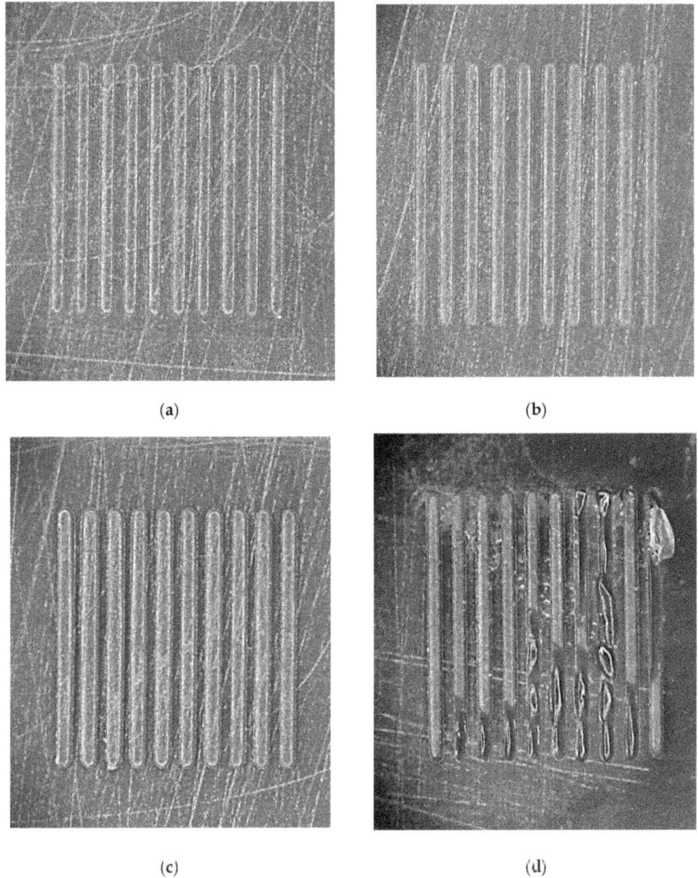

Figure 9. Optical micrographs of intact microfeatures in (**a**) Group A, (**b**) Group B, and (**c**) Group C on Tool 4 after 60 shots, and (**d**) fractured Group D microfeatures on Tool 5 after 78 shots.

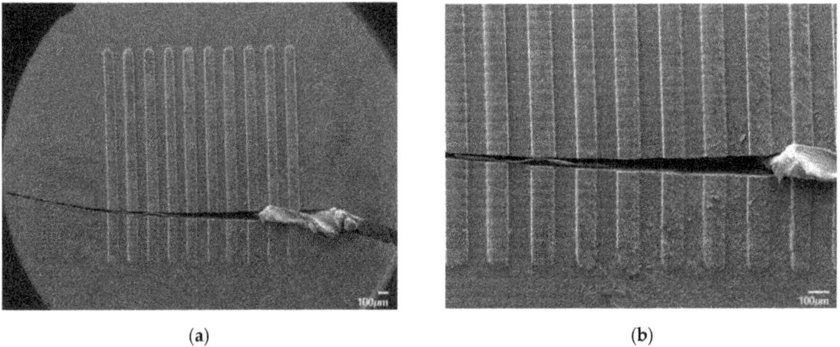

Figure 10. SEM images of Tool 5's Group D channels exhibiting (**a**) brittle failure and (**b**) thermal-induced surface wear.

These results, along with the short tool life for Tools 1–3, indicate that the melt temperature and induced pressure on the tooling can affect the long-term durability of both the bulk insert and the microfeatures on the surface. One way to increase the longevity of higher aspect ratio features would be to metalize the surface or use electroplating or other metallization techniques, which could increase the HDT and the heat transfer at the part surface. Additionally, longer tool life may be possible by utilizing a higher strength, higher HDT photopolymer formulation.

4.5. Microfeature Replication

Replication of polypropylene microfeature dimensions and shape was acquired through contact profilometry. The average depth ratios (DR) for feature Groups A–D were 99.1 ± 0.4%, 97.8 ± 0.7%, 95.4 ± 1.8%, and 98.6 ± 1.0%, respectively. The total average depth ratio (DR) for all feature groupings was 97.7 ± 1.6%, indicating good replication with this tooling. In this case, aspect ratio had little effect on microfeature replication because all features exhibited high depth ratio values. This finding was not unexpected as the features were relatively wide (119 μm) and had relatively low aspect ratios (<1:1). Moreover, the projections in the tooling surfaces did not hinder the flow of the polymer melt.

Contact profilometry traces, shown in Figure 11, illustrate the replication fidelity. The profilometry traces were taken from a cross-section of the parts molded from Tool 4. These traces indicate that the lands of the "channels" (which were molded from the base of the tooling) were flat and at a uniform height. The bottoms of the channels were rounded, reflecting the curvature at the top of the projections in the tooling. The shape of the microfeatures also was consistent over multiple molding cycles. Overall, these characteristics indicate a well-replicated system. It should be noted that the contact profilometry exhibited some limitations in the acquisition of the side walls of the microstructures. Indeed, the vertical resolution does not allow accurate reconstructions of the features, as observed in Figures 11 and 12. This limitation, however, does not affect the vertical resolution of the structures that was used as the response variable for the study (cf. Equation (1)).

While the line scans in Figure 12 were indicative of most measured parts, a few traces suggested damage to the Group A features during ejection. The overall severity of the peaks in these parts varied, as did their existence in general. Prior research found similar defects in molded microfeatures and attributed them to demolding issues where the polymer adheres to the feature wall during part ejection [46,47].

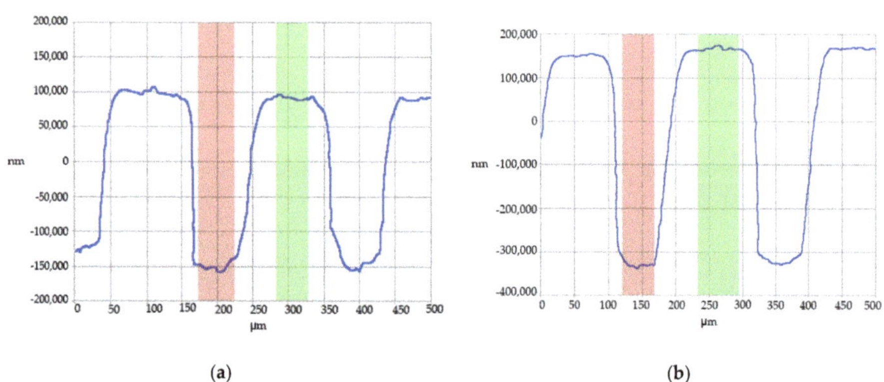

(a)   (b)

Figure 11. *Cont.*

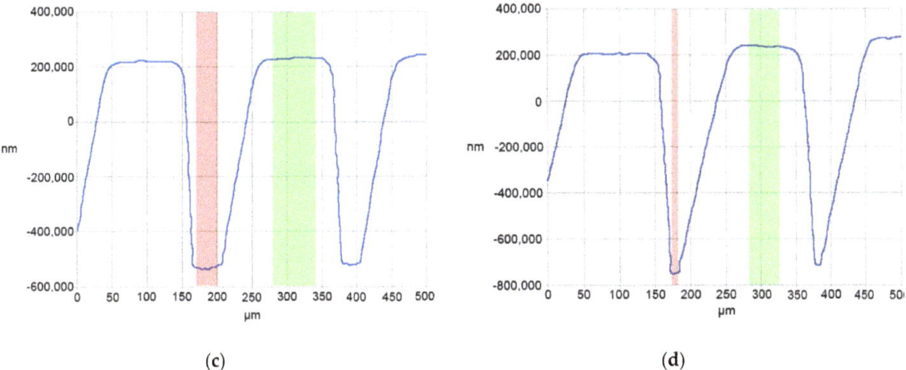

(c) (d)

Figure 11. Contact profilometry traces for (**a**) part 45 from Group A, (**b**) part 30 from Group B, (**c**) part 25 from Group C, and (**d**) part 5 from Group D microfeatures molded from Tool 4.

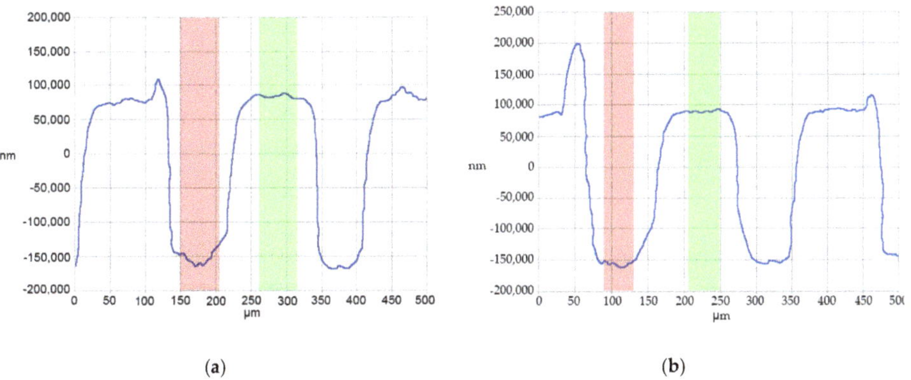

(a) (b)

Figure 12. Contact profilometry traces of damaged Group A channels from Tool 4 parts (**a**) 25 and (**b**) 30.

5. Simulation Results

The numerical model developed in Section 3 was used to study the effect of selecting different mold materials on the injection molding process. For each one of the selected mold materials, a complete Fill + Pack + Cool (Transient) + Warp analysis was run. This approach allowed studying of the transient thermal behavior of the mold. In fact, the thermal analysis is run before the injection starts, in order to determine the steady-state condition, and then during the cycle in order to evaluate the temperature evolution during the process. The temperature of the part and the mold were monitored in the microstructured area for a single cycle, as shown in Figure 13. For each group of microstructures, a probe was inserted to acquire the temperature of the plastic at the entrance of the microstructures. For each group of microstructures in the mold, a sensor was used to acquire the temperature of the tooling at a distance of 0.1 mm from the bottom of each microfeature cavity. This allowed the evaluation of the effect of the aspect ratio on the microscale cavity depth.

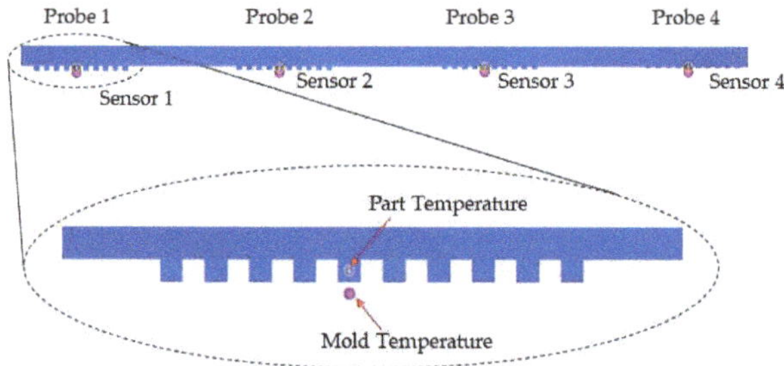

Figure 13. Probe and sensor nodes for monitoring part and mold temperature in the microfeatures.

The results of the analysis are reported in Table 9. Analyzing the part and mold temperature at different stages of the molding process indicated a stable temperature in the part during filling and permitted evaluation of the effect of microstructures' positions during cooling. The results show that the different mold material has a negligible effect on the temperature during filling. The differences, however, became significant when considering the temperature distribution during cooling. The tool temperature during cooling was significantly affected by the mold material. In particular, the low thermal conductivity of the 3D printed tool results in higher part temperatures, leading to more difficult heat transfer to the mold base. The slower cooling with the 3D printed tool also was accentuated by the higher mold temperature, which results in increased wear for the microfeatures in the tool. The temperature distribution was also less homogeneous compared to the aluminum and steel tools, in which the probes have reported similar temperature values.

Table 9. Maximum temperature values for the part and the mold monitored for the different mold materials.

	P20 Steel			Aluminum 6061			3DP Plastic		
	Part		Mold	Part		Mold	Part		Mold
	Fill/Pack	Cooling		Fill/Pack	Cooling		Fill/Pack	Cooling	
A	193.0	82.0	80.1	192.9	75.8	75.4	193.6	110.3	99.4
B	194.7	133.2	83.3	194.5	130.0	76.7	194.9	170.2	101.8
C	195.1	165.0	84.2	195.1	163.5	77.5	195.2	185.3	100.3
D	194.8	86.5	84.0	194.8	78.0	77.1	194.5	101.9	99.8
Average	194.4	116.7	82.9	194.4	111.8	76.7	194.5	141.9	100.3
Dev. Std.	0.9	39.7	1.9	1.0	42.6	0.9	0.7	42.0	1.1
Max.	195.1	165.0	84.2	195.1	163.5	77.5	195.2	185.3	101.8

In the SLA 3D printed plastic, the reduced heat transfer promoted the increase of temperature for microfeatures in groups B and C. These feature groups had significantly higher part temperatures during cooling due to the increased difficulty of removing heat from the center of the part. The higher temperature resulted in higher stiction during the ejection phase and thus in easier breakage of the microstructures in the tool. This breakage was more evident for higher aspect ratio features, which have a higher contact area with the mold surface. When considering the mold temperature, the different aspect ratio had no effect on the mold temperature measured at a distance of 0.1 mm from the bottom of the micro cavity.

Figure 14 shows the history of temperature in the part and the mold, as measured close to the microfeatures. It can be observed that the metal tools are characterized by a significantly smaller increase in part temperature during filling (Figure 14a). Conversely, the temperature of the polymer increases steeply with the 3D printed tool, and it does not recover after filling upon cooling. The effect of the low thermal conductivity is also visible when considering the mold temperature in Figure 14b. Indeed, the plastic tool leads to a stronger rise in the mold temperature, which is not recovered after filling. These results suggest that the short life of the 3D printed tools can be explained the long time required to stabilize the mold temperature after the cycle. The time required to reach the set mold temperature, however, would require a very long time, thus increasing the cycle time significantly. Moreover, the longer time significantly affects the viscosity and stability of the melt (shot) waiting to then be delivered in the next molding cycle.

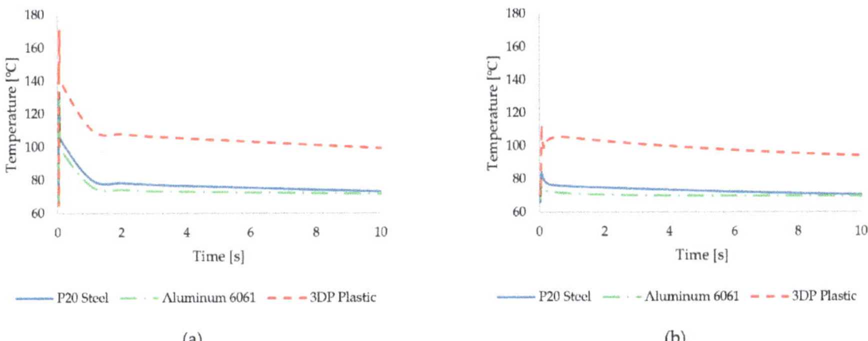

(a)

(b)

Figure 14. Part (**a**) and mold (**b**) temperature probed close to the microfeatures of Group C for the different mold insert materials (cf. Table 7).

The reduced tool life well correlates to the convective thermal load applied by the injected polymer and the reduced thermal conductivity of the insert. A numerical model was created using Moldex3D to analyze the temperature distribution during processing and to compare the 3D printed tool performances with those of conventional metallic mold materials. The simulation results indicated a significant rise in part and mold temperature with the plastic mold. This resulted in reduced heat dissipation to the mold base and eventually in the failure of the soft tooling (cf. Figures 9 and 11). The thermal loads were so significant that failure occurred for the bulk insert, before the micro cavities could be worn off by the polymer flow.

Tooling inserts with projections for simple 100-µm wide microfluidic channels with several aspect ratios were printed on a DLP-SLA system and used to injection mold polypropylene parts. The use of the soft tooling for micro injection molding was analyzed, considering polymer replication and tool life. Injection molding trials and the simulation showed that the DLP-SLA were feasible tooling inserts for injection molding of parts with micro-structured surfaces such as microfluidic channels. Firstly, the DLP-SLA system produced the varying height of tooling geometry to an acceptable deviation from the design, with an overall XY growth of 19% and a depressed Z dimension of 6%. Secondly, the molded parts exhibited high overall replication, with an average depth ratio of 97.7 ± 1.6%. Feature definition also was excellent, with flatlands and rounded channel bottoms in the molded parts as well as relatively smooth part walls. Thirdly, once an optimized injection molding process was established, tooling insert life was 78 shots without bulk failure of the insert. Consistent damage to the tooling projects with the highest aspect ratio (0.79) indicated a minimal aspect ratio for the tooling material. Brittle failure was the dominant failure mode for most inserts.

Modeling of the micro injection molding process with different tool materials allowed for an understanding of the effect of the 3D printed plastic tool. The thermal behavior of the micro-structured tool was observed to be very different compared to conventional metals used for injection molds. The low thermal conductivity of the plastic hinders the heat flow from the part to the mold. This leads to a high thermal load for the microstructures, which fail after repeated cycling. The thermal wear was more accentuated for microstructures with higher aspect ratio, which are more difficult to cool. This indicates the importance of mold design on the wear resistance of the 3D printed tool. Further work will focus on the study of the part/mold thermal boundary condition and its effect on the evolution of the tooling temperature cycle after cycle.

Author Contributions: Conceptualization, D.D. and C.B.; injection molding methodology, D.D. and C.B.; software (simulations), S.M. and D.M.; formal analysis, D.D., S.M., D.M., and C.B.; writing—original draft preparation, D.D., D.M., and C.B.; writing—review and editing, D.M. and C.B.; supervision, D.M. and C.B. All authors have read and agreed to the published version of the manuscript.

Funding: This project received no external funding.

Acknowledgments: The authors would like to thank Nissei America for the micro injection molding machine and Kevin Billett of EnvisionTEC GMBH for the production of the tooling inserts. The authors acknowledge CoreTech System Co., Ltd. for providing the licenses for Moldex3D to the Department of Plastics Engineering at the University of Massachusetts Lowell.

Conflicts of Interest: The authors declare no conflict of interest. The funders had no role in the design of the study; in the collection, analyses, or interpretation of data; in the writing of the manuscript, or in the decision to publish the results.

References

1. Saarikoski, I.; Joki-Korpela, F.; Suvanto, M.; Pakkanen, T.T.; Pakkanen, T.A. Superhydrophobic elastomer surfaces with nanostructured micronails. *Surf. Sci.* **2012**, *606*, 91–98. [CrossRef]
2. Schulz, U. Review of modern techniques to generate antireflective properties on thermoplastic polymers. *Appl. Opt.* **2006**, *45*, 1608–1618. [CrossRef]
3. Mann, E.E.; Mettetal, M.R.; May, R.M.; Drinker, M.C.; Stevenson, B.C.; Baiamonte, V.L.; Marso, J.M.; Dannemiller, E.A.; Parker, A.E.; Reddy, S.T.; et al. Sande, "Surface Micropattern Resists Bacterial Contamination Transferred by Healthcare Practioners". *J. Microbiol. Exp.* **2014**, *1*, 32–39.
4. Sin, M.L.; Gao, J.; Liao, J.C.; Wong, P.K. System Integration—A Major Step Toward Lab on a Chip. *J. Biol. Eng.* **2011**, *5*, 6. [CrossRef] [PubMed]
5. Volpatti, L.R.; Yetisen, A.K. Commercialization of microfluidic devices. *Trends Biotechnol.* **2014**, *32*, 347–350. [CrossRef] [PubMed]
6. Zhang, N.; Srivastava, A.; Kirwan, B.; Byrne, R.; Fang, F.; Browne, D.; Gilchrist, M.D. Manufacturing microstructured tool inserts for the production of polymeric microfluidic devices. *J. Micromech. Microeng.* **2015**, *25*, 95005. [CrossRef]
7. Tsao, C.-W. Polymer Microfluidics: Simple, Low-Cost Fabrication Process Bridging Academic Lab Research to Commercialized Production. *Micromachines* **2016**, *7*, 225. [CrossRef]
8. Masato, D.; Sorgato, M.; Parenti, P.; Annoni, M.; Lucchetta, G. Impact of deep cores surface topography generated by micro milling on the demolding force in micro injection molding. *J. Mater. Process. Technol.* **2017**, *246*, 211–223. [CrossRef]
9. Becker, H.; Gärtner, C. Polymer Microfabrication Methods for Microfluidic Analytical Applications. *Electrophor. Int. J.* **2000**, *21*, 12–26. [CrossRef]
10. Dumond, J.; Low, H.Y. Recent developments and design challenges in continuous roller micro- and nanoimprinting. *J. Vac. Sci. Technol. B* **2012**, *30*, 010801. [CrossRef]
11. Birkar, S.; Mendible, G.; Park, J.-G.; Mead, J.; Johnston, S.P.; Barry, C.M.F. Effect of feature spacing when injection molding parts with microstructured surfaces. *Polym. Eng. Sci.* **2016**, *56*, 1330–1338. [CrossRef]
12. George, N.; Dempsey, D.P.; Yoon, S.-H.; Paszkowski, E.; McGee, M.; Mead, J.L.; Barry, C.M.F. Effect of Polymer Material and Grade When Injection Molding Microfeatures on a Single Platform. In Proceedings of the Annual Technical Conference of Society of Plastics Engineers, Boston, MA, USA, 1–5 May 2011.
13. Heckele, M.; Schomburg, W.K. Review on micro molding of thermoplastic polymers. *J. Micromech. Microeng.* **2003**, *14*, R1–R14. [CrossRef]
14. Attia, U.M.; Marson, S.; Alcock, J. Micro-injection moulding of polymer microfluidic devices. *Microfluid. Nanofluid.* **2009**, *7*, 1–28. [CrossRef]
15. Sha, B.; Dimov, S.; Griffiths, C.; Packianather, M.S. Investigation of micro-injection moulding: Factors affecting the replication quality. *J. Mater. Process. Technol.* **2007**, *183*, 284–296. [CrossRef]

16. Yoon, S.-H.; Cha, N.-G.; Lee, J.S.; Park, J.-G.; Carter, D.J.; Mead, J.; Barry, C.M.F. Effect of processing parameters, antistiction coatings, and polymer type when injection molding microfeatures. *Polym. Eng. Sci.* **2009**, *50*, 411–419. [CrossRef]

17. Hansen, H.N.; Hocken, R.; Tosello, G. Replication of micro and nano surface geometries. *CIRP Ann.* **2011**, *60*, 695–714. [CrossRef]

18. Singh, A.; Metwally, K.; Michel, G.; Queste, S.; Robert, L.; Khan-Malek, C. Injection Moulding Using an Exchangeable Si Mould Insert. *Micro Nanosyst.* **2011**, *3*, 230–235. [CrossRef]

19. Malek, C.K.; Robert, L.; Michel, G.; Singh, A.; Sahli, M.; Manuel, B.G. High resolution thermoplastic rapid manufacturing using injection moulding with SU-8 based silicon tools. *CIRP J. Manuf. Sci. Technol.* **2011**, *4*, 382–390. [CrossRef]

20. Pfleging, W.; Hanemann, T.; Torge, M.; Bernauer, W. Rapid fabrication and replication of metal, ceramic and plastic mould inserts for application in microsystem technologies. *Proc. Inst. Mech. Eng. Part C J. Mech. Eng. Sci.* **2003**, *217*, 53–63. [CrossRef]

21. Han, J.; Han, J.; Lee, B.S.; Lim, J.; Kim, S.M.; Kim, H.; Kang, S. Elimination of Nanovoids Induced During Aspect-Ratio nanostructures by the Pulse reverse Current Electroforming Process. *J. Micromech. Microeng.* **2012**, *22*, 065004. [CrossRef]

22. Parenti, P.; Masato, D.; Sorgato, M.; Lucchetta, G.; Annoni, M. Surface footprint in molds micromilling and effect on part demoldability in micro injection molding. *J. Manuf. Process.* **2017**, *29*, 160–174. [CrossRef]

23. Bissacco, G.; Hansen, H.N.; Tang, P.T.; Fugl, J. Precision Manufacturing Methods of Inserts for Injection Molding of Microfluidic Systems. Proceedings of ASPE Spring Topical Meeting on Precision Micro/Nano Scale Polymer Based Component and Device Fabrication, Columbus, OH, USA, 18–29 April 2005.

24. Sorgato, M.; Masato, D.; Lucchetta, G. Effect of vacuum venting and mold wettability on the replication of micro-structured surfaces. *Microsyst. Technol.* **2016**, *23*, 2543–2552. [CrossRef]

25. Shinde, M.S.; Ashtankar, K.M.; Kuthe, A.M.; Dahake, S.W.; Mawale, M.B. Direct rapid manufacturing of molds with conformal cooling channels. *Rapid Prototyp. J.* **2018**, *24*, 1347–1364. [CrossRef]

26. Rännar, L.-E.; Glad, A.; Gustafson, C.-G. Efficient cooling with tool inserts manufactured by electron beam melting. *Rapid Prototyp. J.* **2007**, *13*, 128–135. [CrossRef]

27. Vaezi, M.; Seitz, H.; Yang, S. A review on 3D micro-additive manufacturing technologies. *Int. J. Adv. Manuf. Technol.* **2012**, *67*, 1721–1754. [CrossRef]

28. Mendible, G.A.; Rulander, J.A.; Johnston, S. Comparative study of rapid and conventional tooling for plastics injection molding. *Rapid Prototyp. J.* **2017**, *23*, 344–352. [CrossRef]

29. Vasco, J.; Pouzada, A. A study on microinjection moulding using moulding blocks by additive micromanufacturing. *Int. J. Adv. Manuf. Technol.* **2013**, *69*, 2293–2299. [CrossRef]

30. Ravi, B. Effects of injection molding parameters on shrinkage and weight of plastic part produced by DMLS mold. *Rapid Prototyp. J.* **2009**, *15*, 179–186. [CrossRef]

31. Sutton, A.T.; Kriewall, C.S.; Leu, M.C.; Newkirk, J.W. Powders for Additive Manufacturing Processes: Characterization Techniques and Effects on Part Properties. In Proceedings of the 27th Annual International Solid Freeform Symposium, Austin, TX, USA, 8–10 August 2016; pp. 1004–1030.

32. Turner, B.N.; Gold, S.A. A review of melt extrusion additive manufacturing processes: II. Materials, dimensional accuracy, and surface roughness. *Rapid Prototyp. J.* **2015**, *21*, 250–261. [CrossRef]

33. Stratasys. Objet Eden260VS. Available online: www.stratasys.com/polyjet-systems (accessed on 18 November 2018).

34. 3D Systems. ProJet 6000HD. Available online: https://www.3dsystems.com/3d-printers/projet-6000-hd (accessed on 18 November 2018).

35. EnvisionTEC. Perfactory 4 Mini. Available online: https://envisiontec.com/3d-printers/perfactory-family/p4 -mini/ (accessed on 18 November 2018).

36. Gheisari, R.; Bártolo, P.; Goddard, N.; Domingos, M.A.N. An experimental study to investigate the micro-stereolithography tools for micro injection molding. *Rapid Prototyp. J.* **2017**, *23*, 711–719. [CrossRef]

37. Masato, D.; Sorgato, M.; Lucchetta, G. Analysis of the influence of part thickness on the replication of micro-structured surfaces by injection molding. *Mater. Des.* **2016**, *95*, 219–224. [CrossRef]

38. Stansbury, J.W.; Idacavage, M.J.; Information, P.E.K.F.C. 3D printing with polymers: Challenges among expanding options and opportunities. *Dent. Mater.* **2016**, *32*, 54–64. [CrossRef] [PubMed]

39. Moldex3D R16 Help, Mathematical Models and Assumptions for standard injection molding. Available online: http://support.moldex3d.com/r16/en/standardinjectionmolding_flow_reference_mathematicalmo delsandassumptions.html (accessed on 21 August 2020).

40. Moldex3D R16 Help, Numerical Method for standard injection molding. Available online: http://supp ort.moldex3d.com/r16/en/standardinjectionmolding_flow_reference_numericalmethod.html (accessed on 12 August 2020).

41. Gibson, I.; Rosen, D.W.; Stucker, B. *Additive Manufacturing Technologies*; Springer: New York, NY, USA, 2010.

42. EnvisionTEC, Advanced DLP for Superior Printing. In *EnvisionTEC White Papers*; EnvisionTEC: Gladbeck, Germany, 2017.

43. Palanisamy, P.; Yoon, S.H.; Lee, J.; Lee, K.H.; Cha, N.G.; Mead, J.L.; Barry, C.M. Enhancement of Surface Replication by Micro-Gas-Injection Molding. In Proceedings of the Annual Technical Conference of Society of Plastics Engineers, Indianapolis, IN, USA, 23–25 May 2009.

44. Palanisamy, P. Enhancement of Surface Replication by Gas Assisted Micro Injection Molding. Master's Thesis, University of Massachusetts Lowell, Lowell, MA, USA, 2009.

45. Masato, D.; Sorgato, M.; Lucchetta, G. Characterization of the micro injection-compression molding process for the replication of high aspect ratio micro-structured surfaces. *Microsyst. Technol.* **2016**, *23*, 3661–3670. [CrossRef]

46. Kim, Y. Investigation of Micro and Nanoscale Surfaces Replication Using Injection Molding. Ph.D. Thesis, University of Massachusetts Lowell, Lowell, MA, USA, 2012.

47. Sorgato, M.; Masato, D.; Lucchetta, G. Effects of machined cavity texture on ejection force in micro injection molding. *Precis. Eng.* **2017**, *50*, 440–448. [CrossRef]

 *micromachines*

Technical Note

3D Printed Reconfigurable Modular Microfluidic System for Generating Gel Microspheres

Xiaojun Chen * , Deyun Mo and Manfeng Gong

School of Mechanical and Electronic Engineering, Lingnan normal university, Zhanjiang 524048, China; deyun_mo@163.com (D.M.); gongmf@lingnan.edu.cn (M.G.)
* Correspondence: chxj@lingnan.edu.cn

Received: 17 January 2020; Accepted: 20 February 2020; Published: 21 February 2020

Abstract: Integrated microfluidic systems afford extensive benefits for chemical and biological fields, yet traditional, monolithic methods of microfabrication restrict the design and assembly of truly complex systems. Here, a simple, reconfigurable and high fluid pressure modular microfluidic system is presented. The screw interconnects reversibly assemble each individual microfluidic module together. Screw connector provided leak-free fluidic communication, which could withstand fluid resistances up to 500 kPa between two interconnected microfluidic modules. A sample library of standardized components and connectors manufactured using 3D printing was developed. The capability for modular microfluidic system was demonstrated by generating sodium alginate gel microspheres. This 3D printed modular microfluidic system makes it possible to meet the needs of the end-user, and can be applied to bioassays, material synthesis, and other applications.

Keywords: modular microfluidic system; 3D printing; gel microspheres

1. Introduction

The reconfigurable microfluidic system are connected by basic module components together to form an integrated microfluidic system, which could achieves specific functions of biochemical analysis, such as emulsion generation [1], multi-organ-chips [2], gradient generation [3], and biochemical analysis [4]. Existing systems often use a monolithic approach, where chemical reactors, sensors, valves, pumps, and detectors are integrated on a single chip. The typical fabrication methods of monolithic microfluidic systems, which include soft lithography, hot embossing, and femtosecond laser writing, are time consuming and often expensive. Additionally, the complicated fabrication processes would require more attention to the quality control [5]. Any part of failure of a monolithic microfluidic system may require rebuilding the entire system, which will result in long development time and incur substantial costs. The modular design approaches are an effective method to address this integration problem. The advantage of a modular system is made of the assembly of basic modular components, and each module can be designed and tested separately before connecting them together to form a larger system.

Modular architecture using prefabricated microfluidic components could be easily assembled, disassembled, reconfigured, and assembled again. Nevertheless, reconfigurable modular microfluidic systems have brought challenges to ensure leak-free fluidic interconnections between connected microfluidic modules after their assembly. Hsieh et al. presented an advanced Lego®-like swappable fluidic module concept to achieve fully portable, disposable fluidic systems [6]. This module design had self-aligning structures on both the male- and female-type Lego®-like for attaining the improved sealing at block junctions. Bhargava et al. demonstrated an approach to microfluidic device design based on discrete elements. These components were connected by a convex block embedded in a concave block to complete the assembly of the entire microfluidic system [7–9]. The aforementioned microfluidic devices were adopted self-aligning structures for attaining the inter-block sealing at block junctions. The instability of the module-to-module fluidic interconnects required to be considered, easily led to fluid leakage, to operate the integrated device under high pressure. In addition, Rheea and Burns proposed a standard set of modular microfluidic assembly block, using pre-fabricated polydimethylsiloxane (PDMS) blocks [10]. This approach required an additional glue-like, UV curable adhesive, as well as PDMS mixture to connect components. Lee et al. proposed an advanced fabrication and assembly method for modular microfluidic devices using rubber O-rings and metal pins interconnects [11]. From the user's point of view, modular microfluidic devices allowed non-expert users to assemble fully customizable microfluidic devices in minutes. However, these module-to-module or world-to-chip fluidic interconnects are required to be strengthened by additional bonding and sealing processes, which will result in a long development time and incur substantial costs.

Recently, 3D printing technology have been applied to fabricate miniaturized and complicated devices, which are of high structural complexity and design flexibility [12–16]. In this work, we propose an advanced reconfigurable modular microfluidic system employing 3D printed modules with assorted channel geometries that can be easily assembled to create complex, modular, and reversible in three dimensions. The microfluidic system consists of two basic functional components: A screw fastener and an assembly module. Each assembly module has an own unique function (such as inlets, outlets, channels, valves, pumps, mixers, and reservoirs) that is connected together and bonded to form a multi-function microfluidic system using screw interconnects. Furthermore, screw connectors are inserted into each microfluidic module's threaded port to eliminate fluid leakage and enable high pressure actuation. It offers a promising way to realize a larger integrated microfluidic system capable of sophisticated functionalities.

2. Materials and Methods

2.1. Materials and Instruments

Ethanol, dye solutions (red and blue) and food grade mineral oil were purchased from Sinopharm Chemical Reagent Co., Ltd. The dye solution was dissolved with deionized (DI) water. Sodium alginate powder and calcium chloride powder were obtained from Sigma-Aldrich Corporation (Guang Dong, China). Materials for the out phase were 2% (*w/w*) calcium chloride. The pre-gel aqueous phases were 2% (*w/w*) sodium alginate. The PDMS was obtained from Dow Corning (Midland, MI, USA). The PDMS was heat cured at a temperature of 100 °C for 20 min in a vacuum drying oven. Fine sandpaper (P1500 and P2000) was purchased from Xiamen Green Reagent Glass Instrument Co., Ltd. (Xiamen, China). An optical microscope (Mitutoyo MF-U, Chongqing Aote Optical Instrument Co., Ltd., Chongqing, China) was used to take images of the mixer. A laser confocal scanning microscope (Produced by Carl Zeiss AG, OLS1200, Carl Zeiss AG, Oberkochen, Germany) was used to measure the surface roughness of model parts. The COLOR intensity was measured using ImageJ. The experimental process was observed by a CCD camera (UI-2250SE-C-HQ, Shanghai Lingliang Optoelectronics Technology Co., Ltd., Shanghai, China).

2.2. Fabrication of Basic Functional Components

To build the 3D models for all of the 3D microfluidic components in this work, we used the computer-aided design (SolidWorks) software. We assembled the 3D models of the ports, components, and systems within the software, and exported the assemblies to the STL format-a standard file type for 3D printers. All modules were directly fabricated by a 3D printer (ProJet®D3510 SD, 3D Systems). This printer is based on the multi jet modeling (MJM) technology using a print head that jets the photopolymer (VisiJet® Crystal, 3D Systems) and the waxy support material (VisiJet® S300, 3D Systems) layer by layer.

The assembly modules were carefully selected from well-known standard component as shown in Figure 1a,b. The assembly modules, which were designed to a standard, cubic geometric footprint (with a size of 10 by 10 by 10 mm^3), comprised functional elements (assembly module) as well as inlet/outlet modules (screw fastener) for world-to-chip fluidic interconnects. The internal channels of these module were designed into a square cross-section with 0.6 mm side length. Each module had different functionality, which acts as basic building units to construct functional microfluidic device. A multi-function microfluidic system was assembled form several modular components (Figure 1c). The screw fastener acted as module-to-module or world-to-chip fluidic interconnects. The spacers were cylindrical with screw pins at opposite sides. The modules were assembled by interlocking screw pins connectors for each module. Following the 3D printing process, the modules were placed in the convection oven at 80 °C to remove the wax layers. When the wax was fully melted, its residue was first removed in a hot oil bath and was then rinsed by ultrasonication in a water bath that contained detergent. The modules were then further rinsed by ultrasonication in a deionized water bath to ensure that the detergent remains were fully removed, and that each procedure was conducted for at least 30 min.

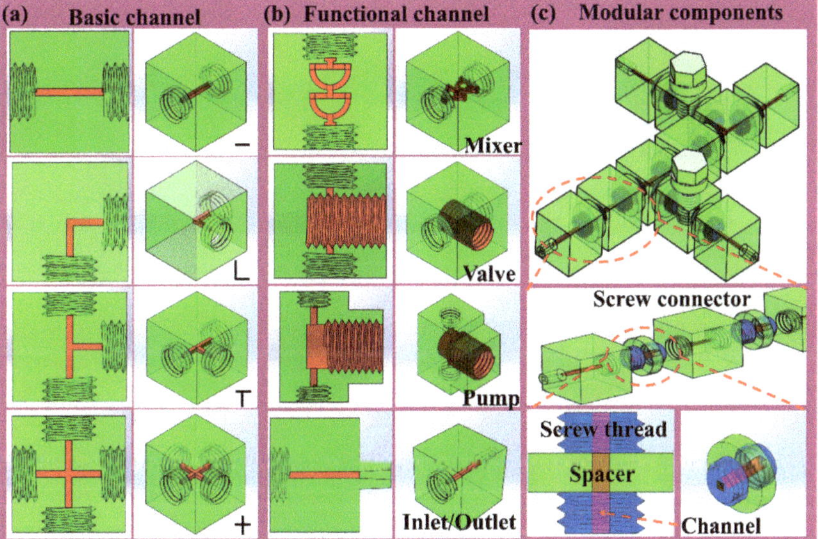

Figure 1. Schematic illustration of 3D printed microfluidic modular components. (**a**) basic microchannel unit. (**b**) functional microchannel unit. (**c**) multifunctional microfluidic system. The device consists of multiple basic channels and functional channels using screw connector.

3. Results and Discussion

3.1. Surface Treatment and Leak Testing

The printed microfluidic channels and their surfaces were observed with an optical microscope, and typical images are shown in Figure 2a. The chip surface quality shown the traces of resin material layers by layers. The surface roughness of the 3D printed module assembly was 2.161 μm. In order to obtain optical transparency, the 3D printed modules were repeatedly polished by two types of fine sandpaper to form a smooth surface. Additionally, the film layer of PDMS on the surface was coated to make the chip more transparent. The surface roughness of the modular component after surface modification was 0.595 μm. Figure 2b demonstrated a modular microfluidic system after polishing and PDMS coating process. The excellent surface quality of the chip was achieved for easier observation of the experiment results. Therefore, surface polishing and PDMS coating were a very effective method to obtain high surface quality. Figure 2c showed that the internal channel structure still appears blurry. An optical microscope described the rough internal flow channel structure. To obtain a highly transparent microchannel, it was also possible to apply PDMS solution inside. Evenly coating the PDMS solution inside the microchannel could enhance the transparency. The key was to prevent the PDMS solution from blocking the internal microchannel.

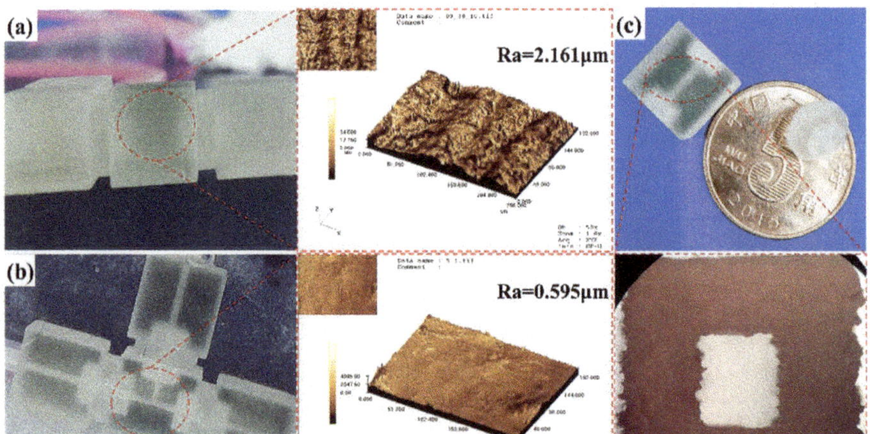

Figure 2. Micrographs of the chip surface quality. (**a**) surface quality and surface roughness before treatment. (**b**) surface quality and surface roughness after polishing and polydimethylsiloxane (PDMS) treatment coating. (**c**) image of 3D printed module components.

Previous literature [17] reported that a wet etching technique was used to treat internal surface roughness of microchannels. Ethanol-diluted acetone solution was circulated in an ultrasonic bath to etch the chip surface. This method significantly reduced the surface roughness of the microchannel without deformation, and the surface roughness after processing was less than 10 nm. Chemical treatment of the internal channels was also an effective method to obtain transparency. Photosensitive resin materials were highly sensitive to organic reagents. Placing the printed module assembly in acetone will dissolve, and high concentrations of methanol and isopropanol (IPA) will whiten the material of the module assembly [18].

To check the capability of screw fastener, we performed three different scenarios such as (a) with using embedded connection [9] (Figure 3a), (b) untightened thread, and (c) tightened thread (Figure 3b). All modules were used digital pressure gauges for leakage tests. The serious level of leakage of bubbles was observed under embedded connection and untightened thread (Figure 3c,d, Movie S1 and S2, Supporting Information). In contrast, the tightened thread modules could be held up to 500 kPa (Figure 3e, Movie S3, Supporting Information). In addition, the Young's modulus of the post-cured clear resin was 2.7 GPa, which is much larger than that of PDMS. Therefore, our devices, printed with this material, were capable of withstanding higher liquid flow rate and input pressure. Compared with the previous literature (Table 1), the method of screw connection was simple and fast. It could be disassembled and reassembled repeatedly to build different functions of novel microfluidic system. It is widely applied in biochemical fields such as fluid mixing and droplet generation. Importantly, the threaded connection provided high fluid pressure, avoiding fluid leakage problems, and expanding its application.

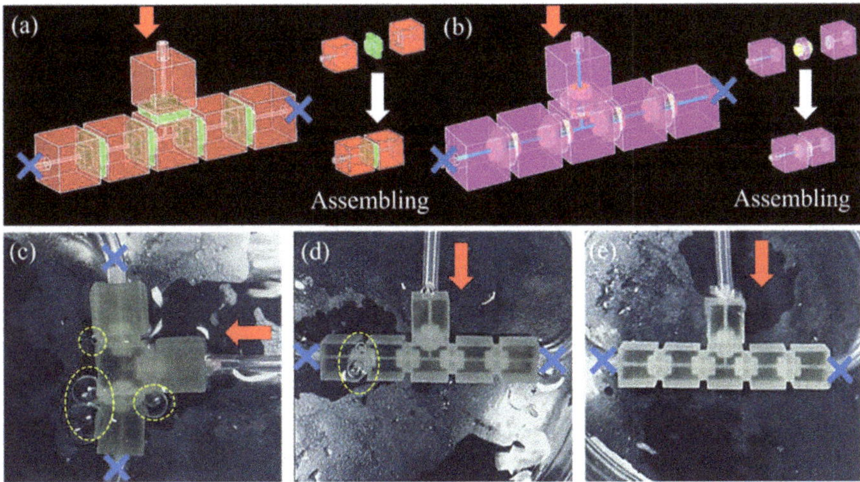

Figure 3. Connectivity of the modular microfluidic system. (**a**) schematic of modular assembly of embedded connections. (**b**) schematic illustration of modular assembly with threaded connections. (**c**) leakage experiment of embedded connected microfluidic system. (**d**) leakage experiment of threaded connected microfluidic system (untightened thread). (**e**) leakage experiment of threaded connected microfluidic system (tightened thread).

Table 1. Comparison of modular microfluidic systems.

References	Connection Type	Manufacturing Characteristics	Fluid Pressure
Bhargava et al. [7–9]	Square interface embedded, comprising spacer and connector	3D printed module, male–male connector aligned with female-type port, reversible strong	As high as 200 mL·h^{-1}
M. Rhee and M. A. Burn [10]	UV-curable glue bonding	PDMS coated glass substrate using the curing agent as the adhesive, reversible weak	40 kPa
Lee et al. [11]	Insert connection, comprising rubber O–ring and metal pins	3D printed module, concave and convex cone–shaped features, require auxiliary components. reversible strong	Up to 200 kPa
Yue [19]	Magnetic interconnects, comprising magnets and sealing gaskets	3D printed module, simple, require auxiliary components	6.8 kPa
Our work	**Screw connector**	**3D printed module, threaded connection, simple, reversible strong**	**Up to 500 kPa**

A reconfigurable modular microfluidic system, comprising a mixer channel and various microfluidic modules as well as inlet/outlet modules for world-to-chip fluidic interconnects, was fabricated and used to demonstrate its reconfigurability to build various integrated microfluidic systems by simply and reversibly assembling various modules together. Three different configurations of a mixer channel modular microfluidic system were built using two (Figure 4a), three (Figure 4b), and four (Figure 4c) outlet channel modules, and two inlet modules. The assembly described in Figure 4a,b were modeled as an equivalent circuit consisting of two branch resistors R ($R = R_{struct} + R_{ref}$) and Rs ($R_s = R_{struct} + R_{select}$) grounded by two water reservoirs and terminated by outlet resistor

Ro (Figure 4d,e). Each branch was designed to differ by only a reference (R_{ref}) and selected (R_{select}) component resistance, while having identical support components resulting in equal structural resistance (R_{struct}). The volumetric mixing ratio M of streams combined in the outlet resistor was predicted by nodal analysis to have simple dependency on only the selected, reference, and branch structural resistances (Equation (1)) [20].

$$M = \frac{R_{struct} + R_{ref}}{R_{struct} + R_{select}} \tag{1}$$

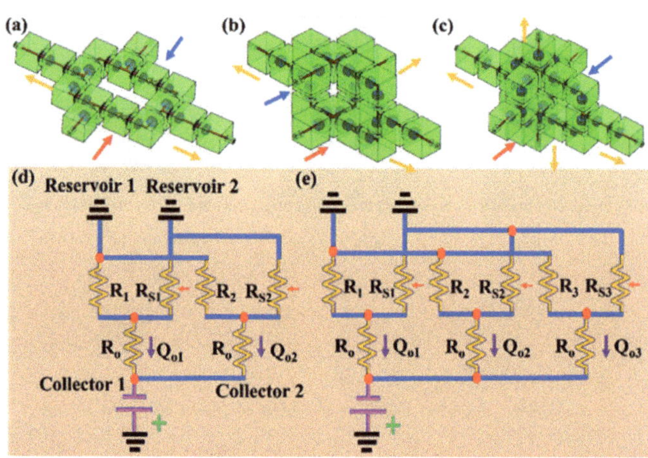

Figure 4. The tunable mixer with (**a**) two, (**b**) three, and (**c**) four outlets. (**d**,**e**) comparison of equivalent circuit models for two- and three-outlet parallelized configurations of the tunable mixer system.

3.2. Reconfigurable Modular System Demonstration

A simple modular microfluidic system was built using one mixer module, a T type module, a flow channel, and three inlet/outlet modules (Figure 5a). To demonstrate the 3D mixing mechanism, numerical simulation using finite element analysis software (COMSOL Multiphysics 5.1) was performed to simulate the mixing effects of microfluidic system with straight-channel modular and a mixer channel modular (Figure 5b). Two streams of solution with a relative species concentration of 1 mol/L and 0 were injected into the mixer module through the inlets; the concentration fields were obtained by solving the incompressible Navier–Stokes and convection diffusion equations in the stationary mode. In Figure 5c, the cross-section (terminal position) of the microfluidic system with the mixer modular shows a uniform distribution of the concentration gradient after passing through 30 mm in length. The normalized concentration shows that the spiral mixer was evenly distributed compared to the straight channel. The microfluidic system with a mixer can achieve 99.7% mixing efficiency at the terminal, while the straight-channel microfluidic system could only achieve 88.3% mixing efficiency. Red and blue color food dye solutions were used to test the modular microfluidic system. Each food dye solution was pumped into the modular microfluidic system at ~20 µL/min by micropump. The dye solution did not have sufficient time to mix through laminar diffusion; hence, the different unmixed streams (a clear fluid interface) could be seen after they pass through the straight-channel (Figure 5e). However, experiments were performed at the same flow rate in a microfluidic system with a mixer (Figure 5d). As the diffusion of dye solution had occurred and the fluids were fully mixed in the designed length. The experimental values well match the simulated ones.

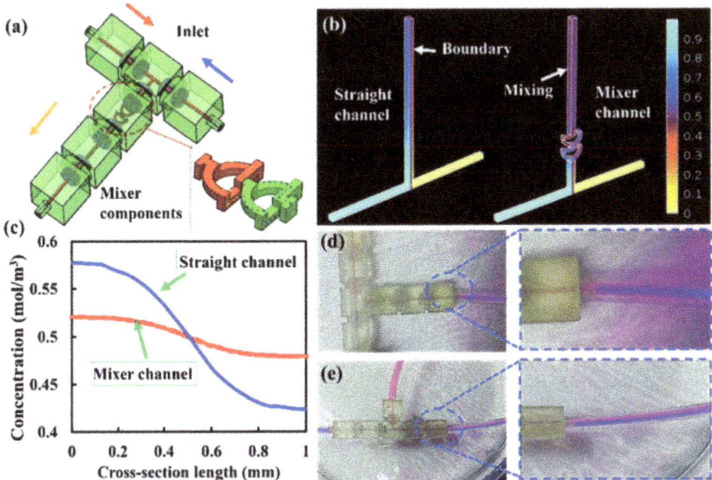

Figure 5. (**a**) schematic illustration of microfluidic system assembly with mixed channel module. (**b**) simulation of mixing effects for straight channel and mixer channel modules. (**c**) normalized concentration of the outlet for the straight channel and mixer channel modules, respectively (the terminal of the microchannel). (**d**,**e**) mixing effects of fluids in straight channel and mixer channel modules.

As a biological hydrogel, sodium alginate was usually encapsulated cells or other biomolecules. Due to its good gelling properties, sodium alginate has been widely used in the field of biomedicine [21]. Sodium alginate was a typical example of a cross-linked hydrogel, which can react with divalent cations to form a hydrogel such as calcium, copper, and iron. To demonstrate the versatility of a modular microfluidic system. A versatile droplet generation microfluidic system had assembled to generate single droplet and dual droplets (Figure 6a). Sodium alginate passed through two branch channels as a carrier phase solution; calcium chloride passed through the main channel as a cross-linking phase solution. Two screw pumps modules in the microfluidic system were used to control the input of two branch solutions. When the sodium alginate was cross-linked with the calcium chloride solution, calcium ions diffused from the outside to the inside, and finally formed the sodium alginate hydrogel (Figure 6b). When one of the threaded pumps in the system was turned off, a single droplet mode was formed. Figure 6c illustrated the formation of sodium alginate droplets (Movie S4, Supporting Information). The formation rate of sodium alginate droplets was mainly related to the diffusion rate, concentration of calcium ions, and the concentration of sodium alginate. When the flow rate of the carrier phase solution was 0.5 mL/h, the relationship between the length, generation frequency of the alginate gel microspheres and the flow rate of the calcium chloride aqueous solution was obtained. As the flow rate of the calcium chloride solution increased, the length of the sodium alginate microspheres decreased. More sodium alginate microsphere gel could be produced as shown in Figure 6d. Finally, the dual droplet generation mode by controlling two screw pump modules was demonstrated (Figure 6e, Movie S5, Supporting Information). The droplet patterns spaced with red and blue can be generated by adjusting the flow rate of the red solution and the blue solution (Figure 6f). Such modular microfluidic systems provided flexibility and versatility to manipulate micro-flows for enhanced and extended applications. Furthermore, the ability to build a reconfigurable modular microfluidic system would be an advantageous platform to substantially enhance design flexibility and improve the system performances of various bio-chemical processes.

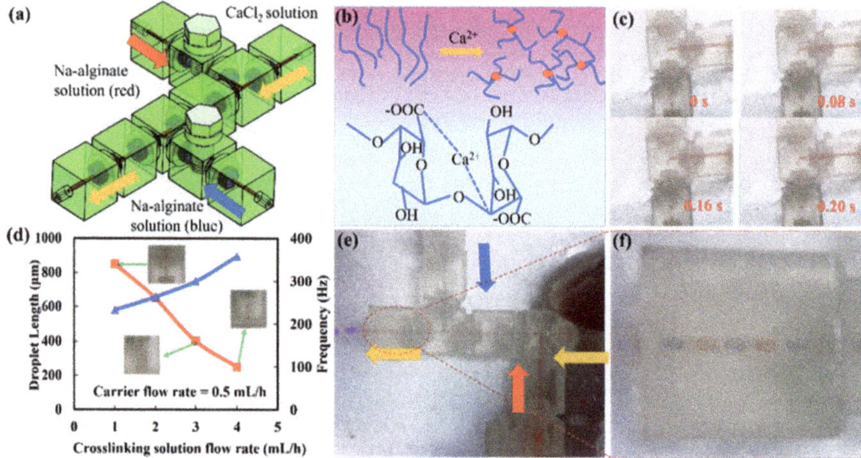

Figure 6. (**a**) modular microfluidic system for droplet generation. (**b**) Ca^{2+} cross-linking to prepare sodium alginate gel microspheres. (**c**) diagram of microsphere generation process. (**d**) the relationship between the length of the gel microspheres and the flow velocity, and the relationship between the flow velocity and the frequency of microsphere generation. (**e**) images of double emulsions. (**f**) Images of double emulsions.

4. Conclusions

In this study, a sample library of standardized components and connectors manufactured using 3D printing was developed. The variety designs of modules were firmly connected using screw fastener. The screw connection perfectly sealed the interconnection between the modules more firmly and prevented any solution leakage. The module assembly and reconstruction are suitable expand microfluidics to non–expert users. Furthermore, the integrated microfluidic device was also applied as multi-analyte mixing and multi-emulsion generation. This modular microfluidic system has a tremendous potential for realizing mass-production complexes and multiplex systems for the commercialization of the microfluidic platform.

Supplementary Materials: The following are available online at http://www.mdpi.com/2072-666X/11/2/224/s1, Movie S1: embedded connection microfluidic system, Movie S2: threaded microfluidic system (untightened), Movie S3: threaded microfluidic system, Movie S4: formation of sodium alginate droplets, Movie S5: dual droplet generation mode.

Author Contributions: Conceptualization, X.C.; Methodology, X.C.; Data curation, X.C.; Writing—original draft preparation, X.C.; Writing—review and editing, X.C., M.G.; visualization, D.M. and X.C.; supervision, D.M., M.G.; project administration, X.C.; funding acquisition, X.C. All authors have read and agreed to the published version of the manuscript.

Funding: This research was funded by the Scientific Research Project Funding of Lingnan Normal University (No. ZL2026).

Conflicts of Interest: The authors declare no conflict of interest.

References

1. Meng, Z.J.; Wang, W.; Liang, X.; Zheng, W.-C.; Deng, N.-N.; Xie, R.; Ju, X.-J.; Liu, Z.; Chu, L.-Y. Plug-n-play microfluidic systems from flexible assembly of glass-based flow-control modules. *Lab Chip* **2015**, *15*, 1869–1878. [CrossRef] [PubMed]

2. Loskill, P.; Marcus, S.G.; Mathur, A.; Reese, W.M.; Healy, K.E. μOrgano: A Lego®-Like Plug & Play System for Modular Multi-Organ-Chips. *PLoS ONE* **2015**, *10*, e0139587. [CrossRef]

3.	Sun, K.; Wang, Z.; Jiang, X. Modular microfluidics for gradient generation. *Lab Chip* **2008**, *8*, 1536–1543. [CrossRef] [PubMed]

4.	Shaikh, K.A.; Ryu, K.S.; Goluch, E.D.; Nam, J.M.; Liu, J.; Thaxton, C.S.; Chiesl, T.N.; Barron, A.E.; Lu, Y.; Mirkin, C.A. A modular microfluidic architecture for integrated biochemical analysis. *Proc. Natl. Acad. Sci. USA* **2005**, *102*, 9745–9750. [CrossRef] [PubMed]

5.	Liou, D.S.; Hsieh, Y.-F.; Kuo, L.-S.; Yang, C.-T.; Chen, P.-H. Modular component design for portable microfluidic devices. *Microfluid. Nanofluid.* **2011**, *10*, 465–474. [CrossRef]

6.	Hsieh, Y.F.; Yang, A.-S.; Chen, J.-W.; Liao, S.-K.; Su, T.-W.; Yeh, S.-H.; Chen, P.-J.; Chen, P.-H. A Lego®-like swappable fluidic module for bio-chem applications. *Sens. Actuat. B Chem.* **2014**, *204*, 489–496. [CrossRef]

7.	Bhargava, K.C.; Thompson, B.; Iqbal, D.; Malmstadt, N. Predicting the behavior of microfluidic circuits made from discrete elements. *Sci. Rep.* **2015**, *5*, 1–9. [CrossRef]

8.	Krisna, B.; Bryant, T.; Anoop, T.; Noah, M. Temperature Sensing in Modular Microfluidic Architectures. *Micromachines* **2016**, *7*, 11–23.

9.	Bhargava, K.C.; Thompson, B.; Malmstadt, N. Discrete elements for 3D microfluidics. *Proc. Natl. Acad. Sci. USA* **2014**, *111*, 15013–15018. [CrossRef]

10.	Rhee, M.; Burns, M.A. Microfluidic assembly blocks. *Lab Chip* **2008**, *8*, 1365–1373. [CrossRef] [PubMed]

11.	Lee, K.G.; Park, K.J.; Seok, S.; Shin, S.; Kim, D.H.; Park, J.Y.; Heo, Y.S.; Lee, S.J.; Lee, T.J. 3D printed modules for integrated microfluidic devices. *RSC Adv.* **2014**, *4*, 32876. [CrossRef]

12.	Farahani, R.D.; Dubé, M.; Therriault, D. Three-Dimensional Printing of Multifunctional Nanocomposites: Manufacturing Techniques and Applications. *Adv. Mater.* **2016**, *28*, 5794–5821. [CrossRef] [PubMed]

13.	Lee, J.M.; Zhang, M.; Yeong, W.Y. Characterization and evaluation of 3D printed microfluidic chip for cell processing. *Microfluid. Nanofluid.* **2016**, *20*, 5. [CrossRef]

14.	Au, A.K.; Huynh, W.; Horowitz, L.F.; Folch, A. 3D-Printed Microfluidics. *Angew. Chem. Int. Ed.* **2016**, *55*, 3862–3881. [CrossRef] [PubMed]

15.	Waheed, S.; Cabot, J.M.; Macdonald, N.P.; Lewis, T.; Guijt, R.M.; Paull, B.; Breadmore, M.C. 3D printed microfluidic devices: Enablers and Barriers. *Lab Chip* **2016**, *16*, 1993. [CrossRef] [PubMed]

16.	Chen, C.; Mehl, B.T.; Munshi, A.S.; Townsend, A.D.; Spence, D.M.; Martin, R.S. 3D-printed microfluidic devices: Fabrication, advantages and limitations—A mini review. *Anal. Method* **2016**, *8*, 6005–6012. [CrossRef] [PubMed]

17.	Wang, Z.K.; Zheng, H.Y.; Lim, R.Y.H.; Wang, Z.F.; Lam, Y.C. Improving surface smoothness of laser-fabricated microchannels for microfluidic application. *J. Micromech. Microeng.* **2011**, *21*, 095008. [CrossRef]

18.	Niall, M.P. Microsystems Manufacturing Technologies for Pharmaceutical Toxicity Testing. Ph.D. Thesis, University of Glasgow, Glasgow, UK, Novermber 2013.

19.	Yuen, P.K. A reconfigurable stick-n-play modular microfluidic system using magnetic interconnects. *Lab Chip* **2016**, *16*, 3700–3707. [CrossRef] [PubMed]

20.	Choi, S.; Lee, M.G.; Park, J.-K. Microfluidic parallel circuit for measurement of hydraulic resistance. *Biomicrofluidics* **2010**, *4*, 034110. [CrossRef] [PubMed]

21.	Cheng, Y.; Zheng, F.; Lu, J.; Shang, L.; Xie, Z.; Zhao, Y.; Chen, Y.; Gu, Z. Bioinspired Multicompartmental Microfibers from Microfluidics. *Adv. Mater.* **2014**, *26*, 5184–5190. [CrossRef] [PubMed]

 *micromachines*

Article

Model for Predicting the Micro-Grinding Force of K9 Glass Based on Material Removal Mechanisms

Hisham Manea [1,2], **Xiang Cheng** [1,*], **Siying Ling** [3], **Guangming Zheng** [1], **Yang Li** [1] and **Xikun Gao** [1]

1 School of Mechanical Engineering, Shandong University of Technology, Zibo 255000, China; hisham.manea@sdut.edu.cn (H.M.); zhengguangming@sdut.edu.cn (G.Z.); liyang0918@163.com (Y.L.); xkg2550@163.com (X.G.)
2 Mechanical Engineering Department, Faculty of Engineering, Sana'a University, Sana'a 12544, Yemen
3 Key Laboratory for Precision & Non-traditional Machining of Ministry of Education, Dalian University of Technology, Dalian 116023, China; lingsy@dlut.edu.cn
* Correspondence: chengxiang@sdut.edu.cn

Received: 11 October 2020; Accepted: 27 October 2020; Published: 29 October 2020

Abstract: K9 optical glass has superb material properties used for various industrial applications. However, the high hardness and low fracture toughness greatly fluctuate the cutting force generated during the grinding process, which are the main factors affecting machining accuracy and surface integrity. With a view to further understand the grinding mechanism of K9 glass and improve the machining quality, a new arithmetical force model and parameter optimization for grinding the K9 glass are introduced in this study. Originally, the grinding force components and the grinding path were analyzed according to the critical depth of plowing, rubbing, and brittle tear. Thereafter, the arithmetical model of grinding force was established based on the geometrical model of a single abrasive grain, taking into account the random distribution of grinding grains, and this fact was considered when establishing the number of active grains participating in cutting $N_{d\text{-}Tot}$. It should be noted that the tool diameter changed with machining, therefore this change was taking into account when building the arithmetical force model during processing as well as the variable value of the maximum chip thickness a_{max} accordingly. Besides, the force analysis recommends how to control the processing parameters to achieve high surface and subsurface quality. Finally, the force model was evaluated by comparing theoretical results with experimental ones. The experimental values of surface grinding forces are in good conformity with the predicted results with changes in the grinding parameters, which proves that the mathematical model is reliable.

Keywords: K9 glass; mathematical model; grinding force; brittle fracture; ductile–brittle transition; active grains number

1. Introduction

Presently, optical glass materials like BK7 or K9 are broadly applied in the production area of various optical accessories for use in visual products or engineering appliances, such as mirrors, prisms, lenses, automobile, aerospace, and panes due to its superior scratch impedance and light transmission properties [1–3]. Therefore, exploring and developing efficient mechanical machining technology of optical glass has become a very important theme. K9 glass is a representative of hard and brittle materials (HBM) [4], which is prone to fracture and damage during machining due to its distinctive features, that is, fragility, resistance, hardness, strength, and alchemical stabilization. As a result, cracks and pits are easily formed on the workpiece surface, affecting the surface quality of the workpiece and the performance of the device [5–7]. Nowadays and because of all these factors, most HBM are processed by grinding due to its high efficiency, but the damage below the surface introduced during grinding has always been a bottleneck problem in machining [8,9]. The most important factor is grinding force, consequently, research on establishing and controlling the grinding force of HBM is particularly important for machining HBM to enhance the grinding efficiency and to upgrade the grinding tool performance [10–13]. The grinding force is closely linked with the quality prediction of the grinding tool surface, geometrical accuracy, and the technique of removing HBM. A lot of studies on micro-grinding force and material removal techniques of HBM have been conducted.

Liu et al. [14] mathematically constructed a force model for the carbon fiber material, assuming that the material removal will behave through a brittle regime, the force model was constructed through the maximum depth of indentation of a single grain in the material workpiece according to the brittle rip theory. A comparable model for K9 optical glass was built by Zhang et al. [15], depended on the mechanism of the indentation rip. The previous scholars have been repealed the forces in the ductile modes as well as neglecting the friction affection in the grinding force. Also, Sun et al. [16] proposed an arithmetical model of the grinding force for Zerodur glass in both brittle and ductile regions, considering the material removal mechanism and the influence of the frictional force. However, this model repealed the fact that not all grits are active during grinding, where the grits are randomly distributed on the wheel with their specific height and width, some of them are active and others have a free ride on the wheel. Moreover, Zhang et al. [17] proposed an arithmetic force model for Silica and Ceramics; the material removal mechanism was separated into two regimes, that are sliding and plowing grinding modes, then the scholars improved the tangential, normal, and radial force models. However, the arithmetic model has not acquired enough attention as the relative motion between the machine tool and the workpiece diverges. Xiao et al. [18] suggested a theoretical model represent the grinding force in the ductile–brittle modes for zirconia material, consideration of brittle–ductile transformation mode, whereas the force model canceled the friction leverage during grinding. Furthermore, Badger and Torrance [19] have been evolved two techniques for predicting the grinding force from wheel surface topography. The first method depends on Chalen and Oxley's 2 D sliding-line domain model of the touch between grinding wheel grain and workpiece surface, while the other method depends on Xie's and Willams 3 D model which creates a chain of channels on the workpiece.

With the concentration on the grinding of glass materials, many researchers have recently suggested mathematical models of grinding force, as well as the study of significant parameters to improve the quality and precision of grinding processes. Chen et al. [20] investigated a reasonable grinding technique for silica glass and acquired the optimal force model identical to the best subsurface fineness; however, the model did not reveal the random distribution of the grinding grains which plays an important role in the model force. Su et al. [21] predicted a model to express the grinding force of silica glass; the model studied how the normal and tangential (NT) forces are affected by variation in the grinding parameters. Then the scholars, after the experimental verification, showed that the force model can represent the grinding force; but the model has not considered the random distribution of the active grits and the changes in the maximum chip thickness accordingly. Zhang et al. [22] kinematically studied the micro-end grinding force of fused silica glass of ultrasonic-assisted through modeling and emulation of abrasive paths. The force model considered the ductile and brittle modes, but the force of the brittle region in the surface-grinding process is yet ambiguous. Consequently, previous studies lacked adequately representative the precise force model of the grinding process for HBM, several important reasons that are the instantaneous variation of grinding space and time as well as the mechanisms of material removal.

In the current article, K9 optical glass is considered a research object to represent the grinding force of HBM. Firstly, theoretical modeling of the micro-grinding force is developed for a single abrasive grit; because of this situation, the maximum chip thickness a_{max} is a dynamic concept in the grinding process, so every single active grain creates a different value of a_{max}. Thereafter, the study constructs the force model in the elastic, plastic, and brittle fracture modes, where the material removal mode is permanent depending on the chip thickness value. The study considered the frictional force generated between the wear protrusion of grits and the grinding face, changes in the value of the frictional coefficient produced by the flux of workpiece over the touch grinding face, the actual active grains which participate in processing, and the reduction in the grinding tool during machining which directly affects the amount of removal chip thickness. Subsequently, the grinding experiments are carried out with a PCD grinding tool and the empirical formulas of NT grinding forces are analyzed; the experimental measurement results are compared to the model prediction values. Finally, the comparative study confirms that the force model can represent the surface grinding force of K9 optical glass.

2. Analysis of Surface Grinding Process

Clarifying the factors of the surface grinding process is a primary task to analyze and correctly establish the grinding force model through the different mechanisms of the material removal modes.

Figure 1 clarifies the mechanism and the main parameters of the radial surface grinding process. The wheel rotates at an angular speed of ω and advances in the direction of the *x*-axis towards the interior surface of the workpiece at a feed rate of f_p.

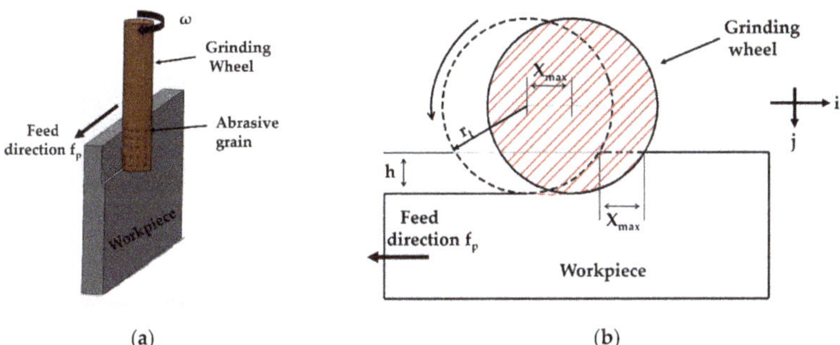

Figure 1. The grinding process: (**a**) Radial surface grinding mechanism; (**b**) grinding zone parameters.

2.1. Displacement and Motion of Single Abrasive Grit

As displayed in Figure 1, the relative displacement of a single grit S with respect to the workpiece is identified as a function of time represented by Equation (1):

$$\vec{S}(t) = \left(r_t \sin(\omega t) + f_p t\right)i + (r_t - r_t \cdot \cos(\omega t))j \tag{1}$$

where r_t is the radius of the grinding tool, f_p is the feed rate, ω is the angular spindle speed, and t defined as the grinding time.

The relative velocity of a single grain to the workpiece V_r is calculated by the first time derivative of the displacement formula as:

$$\vec{V_r}(t) = \frac{\partial S}{\partial t} = \left(r_t \cdot \omega \cdot \cos(\omega t) + f_p\right)i + (r_t \cdot \omega \sin(\omega t))j. \tag{2}$$

2.2. Chip Thickness Model Establishment

The most important step in establishing the force model of the surface grinding process is identifying the regions of the material removal. Therefore, Section 2.2 discusses in detail how to define the various modes of material removal. Figure 2a presents the geometry of the abrasive grain and clarifies the active grain parameters such as semi-apex angle β of the active grain and the average diameter of the active grain D_d which was acquired by gauging the average radius rate of abrasive extremity using AFM.

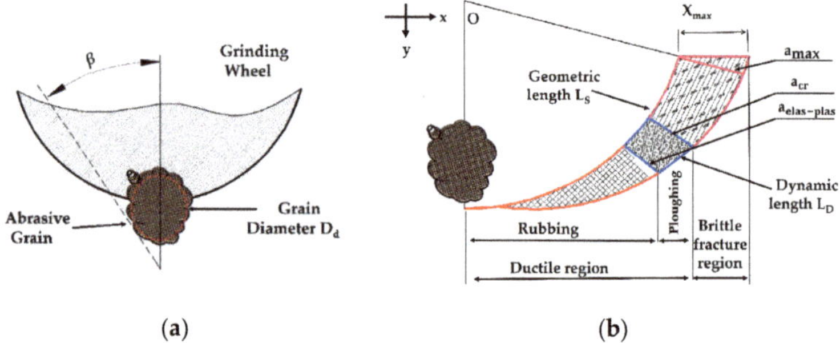

Figure 2. Morphology and motion parameters of abrasive grain: (**a**) Abrasive grain geometry; (**b**) mechanisms of material removal.

As shown in Figure 2b, the entire process of the surface grinding on the interactivity between the grits and the workpiece is classified as ductile region, ductile–brittle transition, and brittle region. The ductile region is divided into three modes, namely, chip formation, elastic, and plastic. Accordingly, the mechanisms of the material removal are classified based on the elastic-to-plastic chip thickness $a_{elas-plas}$, critical chip thickness at ductile-to-brittle transition a_{cr}, and maximum chip thickness a_{max}; can be modeled as follows [23]:

$$\left\{ \begin{array}{r} \left\{ \begin{array}{l} a_c < a_{elas-plas} \text{ (rubbing)} \\ a_{elas-plas} < a_c < a_{cr} \text{ (ploughing)} \end{array} \right\} \text{Ductile region} \\ a_c > a_{cr} \text{ (Brittle region)} \end{array} \right. \tag{3}$$

2.2.1. Chip Thickness of the Elastic-Plastic Transition

Correct modeling of the grinding force needs to consider the recovery and deformation of the elastic mode, in particular at the level of the initial contact zone [23]. So, the designation of the elastic zone needs to locate the transferring from elastic-to-plastic, which is pointed out based on the contact theory of Hertz' and expressed in Equation (4) [24].

$$a_{elas-plas} = \left(\frac{2\pi H_p}{7} \cdot \frac{(1 - \upsilon_p{}^2)}{E_p} \right)^2 \cdot r_d \tag{4}$$

where r_d is the mean radius of the active diamond grain, H_p is the material hardness, υ_p and E_p are the Poisson's ratio and Young's modulus of the workpiece respectively.

2.2.2. Chip Thickness of the Ductile–Brittle Transition

Studying the critical depth of cut a_{cr} is the main key for controlling surface grinding of HBM, as the chip thickness must remain under the a_{cr} to stay grinding in the ductile region [25]. To easily locate the transition point from ductile-to-brittle mode, it is supposed that the grinding parameters do not affect the workpiece specifications during machining, as a result, the critical chip thickness from ductile-to-brittle fracture a_{cr} is expressed by Equation (5) which identified through material properties of the workpiece material [26,27].

$$a_{cr} = \varepsilon \left(\frac{E_p}{H_p} \right) \left(\frac{K_{IC}}{H_p} \right)^2 \tag{5}$$

where ε is a unitless constant ($\varepsilon = 0.15$) [28,29], H_p is the material hardness, E_p is Young's modulus, and K_{IC} tensile strength.

2.2.3. Maximum Chip-Thickness

In the existing study, the maximum chip thickness a_{max} for two uninterrupted grits is established according to the grinding parameters and expressed in Equation (6) [25,30].

$$a_{max} = \left(\frac{3}{4}\right)^{\frac{1}{6}}\left(\frac{\pi}{C_G}\right)^{\frac{1}{3}}\left(\frac{4f_p \cdot r_d{}^2}{0.3\omega \cdot r_t \cdot \tan\beta}\right)^{\frac{1}{2}}\left(\frac{h}{2r_t}\right)^{\frac{1}{4}} = 3.031 \cdot \frac{r_d}{C_G{}^{\frac{1}{3}}} \cdot \left(\frac{f_p}{\omega \cdot \tan\beta} \cdot \sqrt{\frac{h}{r_t{}^3}}\right)^{\frac{1}{2}} \tag{6}$$

where β is defined as the half-oblique angle of a single active grit (suggested as 60° [31]), h is the grinding depth, C_G is the size fracture of grinding grit.

Figure 3a shows the topography of the grinding wheel surface which clearly shows the density of the grinding grains before the grinding process, while Figure 3b shows the effect of grinding on the grains that are dislocated or removed by machining which led to a decrease in the diameter of the grinding wheel accordingly. In general, the diameter of the grinding tool is affected by the high speed, therefore, the diameter decreases by increasing the spindle speed, feed rate, or grinding depth, so the a_{max} is a dynamic concept that can be calculated through Equation (7) [32,33].

$$a_{max} = 3.031 \cdot \frac{r_d}{C_G{}^{\frac{1}{3}}} \cdot \left(\frac{f_p}{\omega \cdot \tan\beta} \cdot \sqrt{\frac{h}{r_t{}^3}}\right)^{\frac{1}{2}} - \frac{X_{max}{}^2}{2r_t}. \tag{7}$$

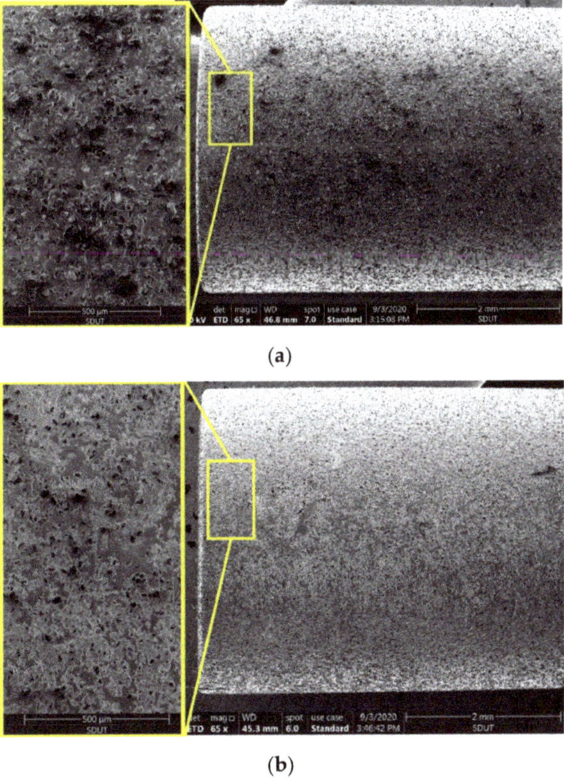

(a)

(b)

Figure 3. SEM captures clarify the wheel surface topography: (**a**) Unused grinding wheel; (**b**) after machining.

The cutting path of the cutting edge of the previous grain moves from the first touchpoint to the workpiece surface. The linear distance from the first to the last touchpoints is equal to the amount of translation of the workpiece X_{max} during the interval between the turning of adjacent cutting edges as displayed in Figure 1b, X_{max} can be calculated by Equation (8).

$$X_{max} = \frac{\lambda_r \cdot f_p}{\omega r_t} \tag{8}$$

where λ_r is the average space between any two active grains calculated by Equation (9) [31].

$$\lambda_r = \left(\sqrt{\frac{\pi}{C_G}} - 2 \right) \cdot r_d. \tag{9}$$

By substitution Equations (8) and (9) into Equation (7), the a_{max} is expressed as

$$a_{max} = 3.031 \cdot \frac{r_d}{C_G^{\frac{1}{3}}} \cdot \left(\frac{f_p}{\omega \cdot \tan \beta} \cdot \sqrt{\frac{h}{r_t^3}} \right)^{\frac{1}{2}} - \left(\sqrt{\frac{\pi}{C_G}} - 2 \right)^2 \cdot \left(\frac{r_d^2}{2r_t} \right) \cdot \left(\frac{f_p}{\omega r_t} \right)^2. \tag{10}$$

2.3. Clarifying the Various Grinding Areas

Identification of the NT grinding areas of a single grain, whether projected or contacting, is a very important task to accurately establish the NT grinding forces of a single grit. The touch depth is almost equivalent to the semi-value of the depth of cut when the grit begins to touch the working surface in the elastic region [24]. Meanwhile, the touch depth is almost equivalent to the depth of cut during plastic mode [34].

As shown in Figure 4, the NT areas depend on the half-vertex angle β of a single grit, the grain diameter D_d and cutting depth a_c or h_t. Equation (11) represents the projection area in normal plane $A_{n\text{-elas}}$, the area in the thrust axis $A_{t\text{-elas}}$ is displayed in Equation (12).

$$A_{n\text{-elas}} = r_d^2 (\beta - 0.5 \sin \beta) \tag{11}$$

where h_t is the cutting depth in the elastic zone and established as $\left(h_t = \frac{a_{max}}{2N_{d\text{-x}}} \right)$ ($N_{d\text{-x}}$ is the active grains through x-axis which is discussed in Section 3.1).

$$A_{t\text{-elas}} = \left(r_d^2 \cdot \beta \right) - \left((r_d - h_t) \cdot r_d \cdot \sin \beta \right). \tag{12}$$

The angle β is set as a dynamic value calculated based on the grinding modes, it, therefore, depends entirely on the cutting depth. In the elastic mode, the angle β is set as $\left(\beta = \cos^{-1} \left(\frac{D_d - 2h_t}{D_d} \right) \right)$.

In the plastic region, the projected areas whether in the normal axis $A_{n\text{-plas}}$ or the tangential axis $A_{t\text{-plas}}$ are established as:

$$A_{n\text{-plas}} = \frac{\pi}{2} \left(2r_d \cdot a_c - a_c^2 \right) \tag{13}$$

$$A_{t\text{-plas}} = r_d^2 \cdot \beta - (r_d - a_c) \cdot \sqrt{2r_d \cdot a_c - a_c^2} \tag{14}$$

where the angle β under plastic mode is calculated as $\left(\beta = \cos^{-1} \left(\frac{D_d - 2a_c}{D_d} \right) \right)$.

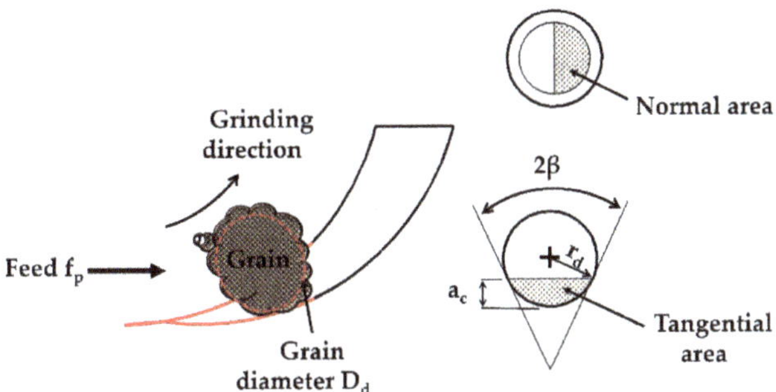

Figure 4. Elastic and plastic projected areas of a single active grain.

3. Grinding Force Establishment

Based on the preceding analyses of the material removal, the force model is classified according to the classification of the grinding modes which stated in Equation (3). The sophisticated force pattern when surface grinding can be determined in two parts, namely, NT grinding forces. The grinding force model is constructed according to the individual grain interactivity, thereupon the formula of the active grains should be established initially. Section 3.1 briefly explains how to identify the number of active diamonds.

3.1. Active Abrasive Grain

The maximum chip thickness a_{max} is a dynamic concept in the grinding process and every single active grit creates a different value based on the wheel properties and grinding parameters as displayed in Figure 5. For this, not all the grains making protrusion in the surface of the grinding tool will engage in the grinding process. So, the number of active diamonds is calculated separately in the $(x$–$z)$ directions, then the total active grits are obtained by multiplying the active grains on x-axis and z-axis. The number of active diamonds formula is established through identifying the touch arc length λ_c as expressed in Equation (15), and the average distance between two active grains λ_r is displayed in Equation (9). The touch arc length λ_c is defined as [34,35]:

$$\lambda_c = \left(1 + \frac{f_p}{V_r}\right) \cdot \sqrt{2\,h \cdot r_t}. \tag{15}$$

The active grains in the x, z-axes are explained in the following equations:

$$N_{d\text{-}x} = \frac{\lambda_c}{2r_d + \lambda_r} = \left(1 + \frac{f_p}{\omega \cdot r_t}\right) \cdot \sqrt{\frac{2C_G \cdot h \cdot r_t}{\pi r_d{}^2}} \tag{16}$$

$$N_{d\text{-}z} = \frac{a_w}{2r_d + \lambda_r} = \frac{a_w}{r_d} \cdot \sqrt{\frac{C_G}{\pi}}. \tag{17}$$

The total active grains $N_{d\text{-}Tot}$ is calculated through multiplying the active grains in x-axis $N_{d\text{-}x}$ and the active grains in z-axis $N_{d\text{-}z}$ as:

$$N_{d\text{-}Tot} = N_{d\text{-}x} \cdot N_{d\text{-}z} = \left(1 + \frac{f_p}{V_r}\right) \cdot \left(\frac{C_G \cdot a_w}{\pi r_d{}^2}\right) \cdot \sqrt{2\,h \cdot r_t}. \tag{18}$$

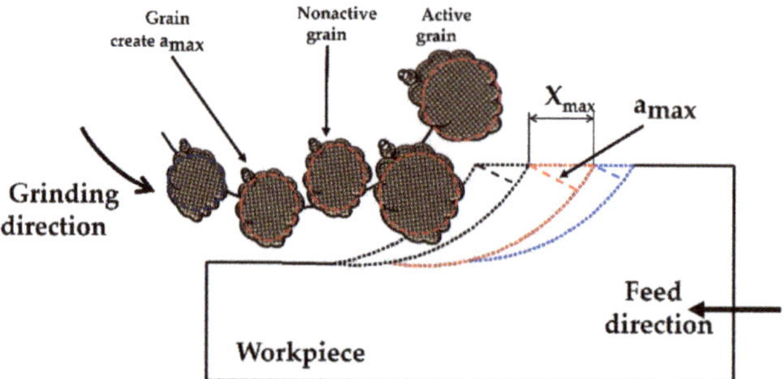

Figure 5. The trajectory and allocation of abrasive grains.

3.2. Force Model in Ductile Region

As shown in Figure 6, the mode of a ductile zone is classified into two categories; elastic and plastic modes. So, the force model taking the frictional force into a consideration is built according to these modes.

$$\begin{cases} F_n = F_{n\text{-elas}} + F_{n\text{-plas}} + F_{n.fr} \\ F_t = F_{t\text{-elas}} + F_{t\text{-plas}} + F_{t.fr} \end{cases} \tag{19}$$

where F_n and F_t represent the total grinding force of NT coordinates respectively, $F_{n\text{-elas}}$ and $F_{t\text{-elas}}$ are the NT rubbing forces acted in the elastic region, $F_{n\text{-elas}}$ and $F_{t\text{-elas}}$ are the NT plowing forces acted in the plastic region, $F_{n.fr}$ and $F_{t.fr}$ are the NT frictional forces which acted through the whole grinding process.

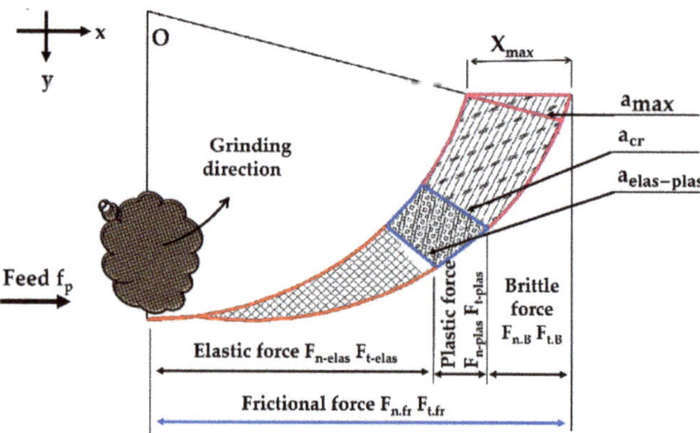

Figure 6. Schematic diagram of grinding force variation during grinding.

3.2.1. Elastic Force

When $a_c < a_{elas\text{-plas}}$ the elastic stage is dominant, the grinding force is named according to the stage name and expressed as follows:

$$F_{n\text{-elas}} = N_{d\text{-Tot}} \cdot P_{max\text{-elas}} \cdot A_{n\text{-elas}} \tag{20}$$

$$F_{t\text{-elas}} = \mu N_{d\text{-Tot}} \cdot P_{max\text{-elas}} \cdot A_{t\text{-elas}} \tag{21}$$

where $P_{max\text{-elas}}$ is the maximum contact pressure occurred in the elastic stage, μ is the coefficient of friction.

To calculate the maximum pressure in elastic and plastic zones, the maximum pressure acted in the ductile mode P_{max} must be clarified initially, therefore it is displayed by Equation (22) [36].

$$P_{max} = \left(\eta + \zeta\left(\frac{\omega \cdot r_t}{R_c \cdot \varepsilon_c}\right)\right) \cdot P_i \tag{22}$$

where η, ζ are constants defined according to the material sensitivity to strain average. ε_c is strain average constant calculated as $\varepsilon_c = 1$ [37], P_i is the contact pressure related to the pressure in elastic zone $P_{i\text{-elas}}$ classified by Equation (24), as well as pressure in the plastic zone $P_{i\text{-plas}}$ expressed by Equation (31), R_c is the contact arc length identified by Equation (23).

$$R_c = 2\left(a_c \cdot D_d - a_c^2\right)^{\frac{1}{2}} \tag{23}$$

where D_d is the average median diameter of a single abrasive grit.

The pressure in the elastic mode can be classified as [24,38]:

$$P_{i\text{-elas}} = \frac{4 \sqrt{2}E_{ef}}{3\pi} \cdot \sqrt{\frac{a_c}{D_d}} \tag{24}$$

where the effective elastic modulus E_{ef} can be calculated through Equation (25).

$$E_{ef} = \left(\frac{1 - \upsilon_p^2}{E_p} + \frac{1 - \upsilon_d^2}{E_d}\right)^{-1} \tag{25}$$

where υ_p, υ_d are the Poisson's ratio of the workpiece and diamond grain respectively, E_p, E_d are the elastic modulus of the workpiece and diamond grain respectively.

Then the maximum pressure in elastic mode $P_{max\text{-elas}}$ can be obtained as follows:

$$P_{max\text{-elas}} = \frac{4 \sqrt{2}E_{ef}}{3\pi} \cdot \left(\eta + \zeta\left(\frac{\omega \cdot r_t}{R_c \cdot \varepsilon_c}\right)\right) \cdot \sqrt{\frac{a_c}{D_d}} \cdot \tag{26}$$

The NT forces that occurred in elastic mode $F_{n\text{-elas}}$ and $F_{t\text{-elas}}$ are respectively expressed by Equations (27) and (28).

$$F_{n\text{-elas}} = N_{d\text{-Tot}} \cdot \frac{2 \sqrt{2}E_{ef}}{3} \cdot \left(\eta + \zeta\left(\frac{\omega \cdot r_t}{R_c \cdot \varepsilon_c}\right)\right) \cdot \sqrt{\frac{a_c}{D_d}} \cdot \left(2r_d \cdot h_t - h_t^2\right) \tag{27}$$

$$F_{t\text{-elas}} = \mu N_{d\text{-Tot}} \cdot \frac{4 \sqrt{2}E_{ef}}{3\pi} \cdot \left(\eta + \zeta\left(\frac{\omega \cdot r_t}{R_c \cdot \varepsilon_c}\right)\right) \cdot \sqrt{\frac{a_c}{D_d}} \cdot \left(\left(r_d^2 \cdot \beta\right) - \left((r_d - h_t) \cdot r_d \cdot \sin\beta\right)\right). \tag{28}$$

3.2.2. Plastic Force

The plastic zone is dominant whenever the $a_{elas\text{-plas}} < a_c < a_{cr}$. The plowing force is created whenever the chip thickness is analogous to the rim of the active grits radius [39]. Zhang et al. [24] recommended that the ploughing force is fundamentally created due to the plastic deformation in the grinding zone of the workpiece. Accordingly, the contact force due to contact pressure represents the plowing force as long as the plastic mode is dominant.

$$F_{n\text{-plas}} = N_{d\text{-Tot}} \cdot P_{max\text{-plas}} \cdot A_{n\text{-plas}} \tag{29}$$

$$F_{t\text{-plas}} = \mu N_{d\text{-Tot}} \cdot P_{max\text{-plas}} \cdot A_{t\text{-plas}}. \tag{30}$$

The contact pressure in the plastic mode $P_{i\text{-plas}}$ is defined in Equation (31) [40].

$$P_{i\text{-plas}} = \mathcal{E}\left(\frac{H_p{}^4}{E_p}\right)^{1/3} \tag{31}$$

where $\mathcal{E}$ is a unitless factor assumed as $\mathcal{E} = 1.5$ for HBM [40].

Then, the maximum pressure in the elastic mode $P_{\max\text{-plas}}$ is calculated as:

$$P_{\max\text{-plas}} = \mathcal{E}\left(\eta + \zeta\left(\frac{\omega \cdot r_t}{R_c \cdot \varepsilon_c}\right)\right)\cdot\left(\frac{H_p{}^4}{E_p}\right)^{1/3}. \tag{32}$$

Consequently, the NT forces created in the plastic zone are identified through Equations (33) and (34) respectively.

$$F_{n\text{-plas}} = \frac{\pi\mathcal{E}}{2}N_{d\text{-Tot}}\left(\eta + \zeta\left(\frac{\omega \cdot r_t}{R_c \cdot \varepsilon_c}\right)\right)\cdot\left(\frac{H_p{}^4}{E_p}\right)^{1/3}\cdot\left(2r_d \cdot a_c - a_c{}^2\right) \tag{33}$$

$$F_{t\text{-plas}} = \mu\mathcal{E}N_{d\text{-Tot}}\left(\eta + \zeta\left(\frac{\omega \cdot r_t}{R_c \cdot \varepsilon_c}\right)\right)\cdot\left(\frac{H_p{}^4}{E_p}\right)^{1/3}\cdot\left(r_d{}^2 \cdot \beta - (r_d - a_c)\cdot\sqrt{2r_d \cdot a_c - a_c{}^2}\right). \tag{34}$$

3.3. Force Model in the Brittle Region

$$\begin{cases} F_n = F_{n\text{-elas}} + F_{n\text{-plas}} + F_{n.fr} + F_{n.B} \\ F_t = F_{t\text{-elas}} + F_{t\text{-plas}} + F_{t.fr} + F_{t.B} \end{cases} \tag{35}$$

where $F_{n.B}$ and $F_{t.B}$ are the NT grinding forces that occurred in the brittle tear region defined through Equations (36) and (37) respectively.

$$F_{n.B} = N_{d\ Tot}\cdot N \cdot T_{ef}\cdot F_{B.max} \tag{36}$$

$$F_{t.B} = \mu F_{n.B} \tag{37}$$

where N is the rotational speed of the spindle, T_{ef} is the effective grinding time of single grinding grain for one cycle expressed in Equation (38), $F_{B.max}$ is the maximum force in the brittle region.

$$T_{ef} = \frac{L_{ef}}{V_r} \tag{38}$$

that L_{ef} is the effective length between the grinding rod and grinding surface expressed as:

$$L_{ef} = \frac{L_D \cdot L_s}{L_D + L_s} \tag{39}$$

where L_S and L_D are the geometric and dynamic contact lengths between the grinding tool and workpiece grinding surface as shown in Figure 2 and modeled by Equations (40) and (41) respectively.

$$L_S = r_t \cos^{-1}\left(\frac{r_t - h}{r_h}\right) \tag{40}$$

$$L_D = \int_0^{T_{Tot}} \sqrt{V_P{}^2 + (2\,r_t\,\omega\,V_P \cos(wt))^2 + (r_t\,\omega)^2}\,dt \tag{41}$$

where T_{Tot} is the total grinding time for one active grit to intersect the contact length L_D calculated as:

$$T_{Tot} = \frac{\cos^{-1}\left(\frac{r_t - h}{r_t}\right)}{2\pi N}.$$ (42)

The maximum force created in the brittle zone is classified by Equation (43) [41].

$$F_{B.max} = \sqrt{\frac{8D_d \cdot a_{max}{}^3}{9} \cdot \left(\frac{E_p}{1 - \upsilon_p{}^2}\right)^2} = \frac{2E_p}{3\left(1 - \upsilon_p{}^2\right)} \cdot \sqrt{2D_d \cdot a_{max}{}^3}.$$ (43)

By substituting Equations (38) and (43) into Equations (36) and (37), $F_{n.B}$ can be obtained as:

$$F_{n.B} = N_{d\text{-}Tot} \cdot \frac{2E_p \cdot N \cdot L_D \cdot L_s}{3V_r\left(1 - \upsilon_p{}^2\right) \cdot (L_D + L_s)} \cdot \sqrt{2D_d \cdot a_{max}{}^3}.$$ (44)

3.4. Frictional Force

In surface grinding of the HBM, the frictional effects are created due to two resources between the abrasives and the grinding surface, one of them is produced by the flux of workpiece over the touch face where the other is generated from the frictional between the wear protrusion of grits and the grinding face. The influence of grinding heat affects the coefficient of friction to vary with varying conditions and behavior of surface grinding. Therefore, modeling the coefficient of friction considering all of these factors is a highly significant task.

$$F_{n.fr} = N_{d\text{-}Tot} \cdot P_{cont} \cdot A_{cont}$$ (45)

$$F_{t.fr} = \mu F_{n.fr}$$ (46)

where P_{cont} and A_{cont} are the contact pressure and contact area between the wear protrusion of active grits and the grinding face respectively.

Through the convergence of the parabolic function with the cutting path, the perversion is located between the arc of the grinding channel and the radius of the grinding wheel and can be calculated by Equation (47) [42].

$$\varrho = \frac{2f_p}{r_t \cdot V_r}.$$ (47)

In the formula, ϱ is the deviation between the arc of the grinding channel and the radius of the grinding rod.

The average touch pressure between a single grit and workpiece P_{cont} is expressed by Equation (48) [43]:

$$P_{cont} = \varrho P_0 = \frac{2P_0 \cdot f_p}{r_t \cdot V_r}$$ (48)

where P_0 is defined as the experiential suitability factor calculated via the experiments.

The frictional coefficient μ is clarified as [44].

$$\mu = \frac{\alpha_1}{P_{cont}} + \alpha_2.$$ (49)

That α_1, α_2 represent the coefficients of the materiality of the frictional pairs.

$$A_{cont} = \int_0^{r_d} (2\beta \, r_d) dr_d - (r_d \sin \beta) \cdot (r_d \cos \beta) = \frac{1}{8} D_d{}^2 \cdot (2\beta - \sin 2\beta).$$ (50)

By substituting Equations (48) and (50) into Equations (45) and (46), $F_{n.fr}$ can be calculated as:

$$F_{n.fr} = N_{d\text{-Tot}} \cdot \frac{P_0 \cdot f_p \cdot D_d^2}{4 r_t \cdot V_r} \cdot (2\beta - \sin 2\beta). \tag{51}$$

In our study, the coefficient of friction is calculated through Equation (52) [45].

$$\mu = 0.3885 - 0.011 \ln(1000\ V_r). \tag{52}$$

4. Predicted Force Model Analysis

In the preceding parts, the arithmetic model of surface grinding force for HBM has been suggested, whilst in the current part, the effects of grinding conditions on the grinding force will be studied through the suggested model.

Surface grinding parameters are listed in Table 1 for grinding force predictions. The relationship between grinding conditions (f_p, a_w, N, h) and NT grinding forces are displayed in Figure 7 according to the derived equations. The average amplitude of the NT forces is almost the same. By observing the force signal curves of the grinding process, it is found that both are the same. This is mainly because of feed speed is very low compared to the rotational speed of the abrasive particles, therefore it has little effect on the grinding force. As shown in Figure 7, the curves of the NT forces are varying with machining parameters during the grinding process. When the abrasive grain maintains its original shape, as the grinding depth h increases, the grinding force increases monotonically, and the NT forces change roughly the same; on the other hand, as the grinding speed increases, the grinding force gradually and nonlinearly decreases. This is consistent with the conclusion that the grinding speed is increased and the cutting force of a single abrasive particle is reduced in the current research on high-speed grinding processing [46]. In short, both NT grinding forces increase with the ascending of grinding conditions of feed rate f_p, grinding depth h, and grinding width a_w; however, decreasing with increasing of spindle speed.

Table 1. Surface grinding conditions for predicted grinding force.

Group No	Feed Rate f_p (mm/min)	Spindle Speed N (min^{-1}) $\times 10^3$	Grinding Depth h (μm)	Grinding Width a_w (mm)	Wheel Diameter D_t (mm)
1	60, 100, 150, 200	10	10	1, 2.5, 5, 7.5 10, 15	4
2	60, 100, 150, 200	5, 7.5, 10, 12.5, 15, 20	10	5	4
3	60, 100, 150, 200	10	5, 7.5, 10, 12.5, 15, 17.5, 20	5	4
4	60, 100, 150, 200, 250	5, 7.5, 10, 12.5, 15, 20	10	5	4

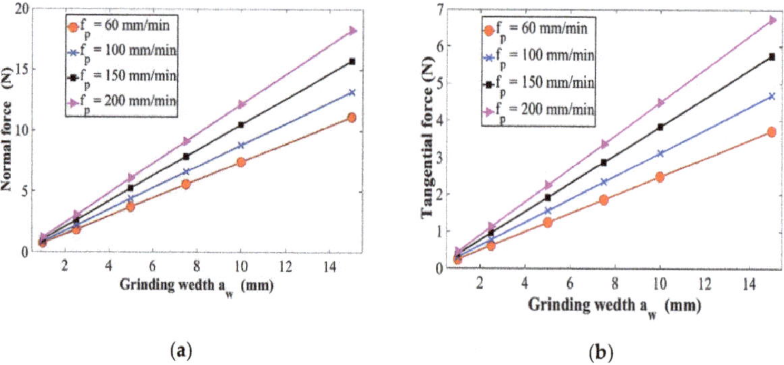

Figure 7. *Cont.*

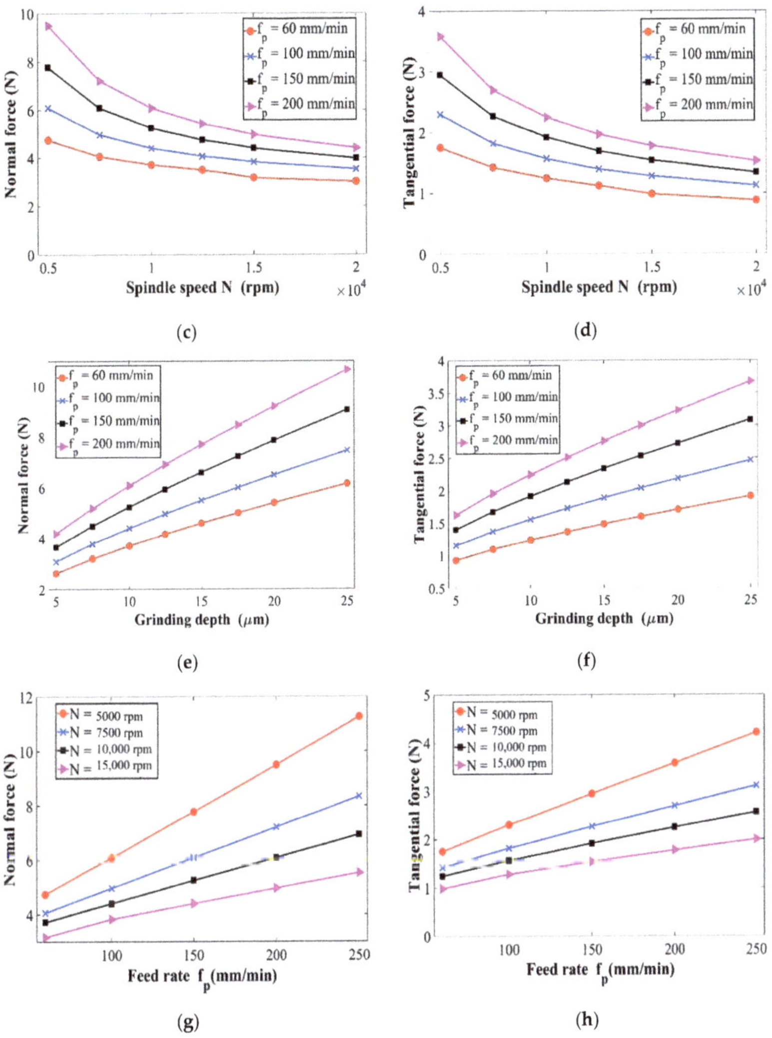

Figure 7. Calculated grinding forces (**a**) $N = 10^4$ min^{-1}, $h = 10$ μm, $r_t = 2$ mm; (**b**) $N = 10^4$ min^{-1}, $h = 10$ μm, $r_t = 2$ mm; (**c**) $h = 10$ μm, $a_w = 5$ mm, $r_t = 2$ mm; (**d**) $h = 10$ μm, $a_w = 5$ mm, $r_t = 2$ mm; (**e**) $N = 10^4$ min^{-1}, $a_w = 5$ mm, $r_t = 2$ mm; (**f**) $N = 10^4$ min^{-1}, $a_w = 5$ mm, $r_t = 2$ mm; (**g**) $h = 10$ μm, $a_w = 5$ mm, $r_t = 2$ mm; (**h**) $h = 10$ μm, $a_w = 5$ mm, $r_t = 2$ mm.

5. Experimental Validation

5.1. Experimental Setup

To validate the arithmetical force model, the experiments were carried out on the precision surface grinder CarverPMS23_A8 (Figure 8). In the experiments, the substrate rotates at a certain speed to realize the grinding of a single abrasive grit where the maximum experimental speed of the spindle was 1.5×10^4 min^{-1}, and the spindle drives the grinding wheel to move longitudinally to adjust the feed rate. The NT forces data is obtained using a high resolution, high precision dynamometer.

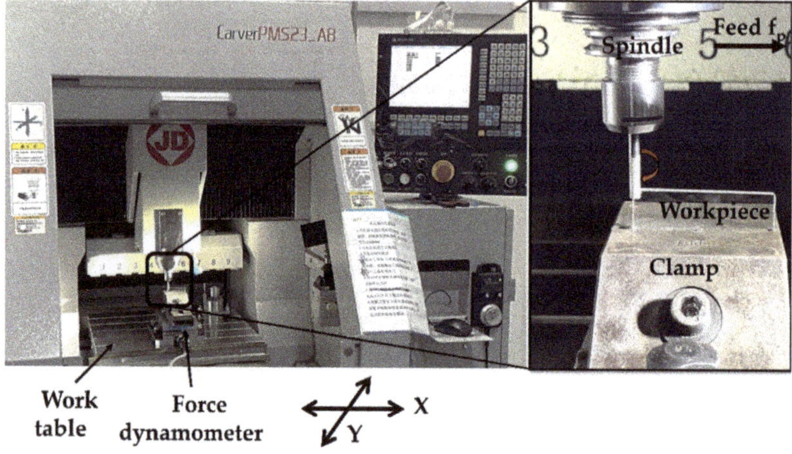

Figure 8. The precision surface grinder CarverPMS23_A8.

The workpiece material used is K9 glass, with the dimensions of 50 mm × 20 mm × 3 mm. Table 2 lists the K9 glass ingredients where its material properties are listed in Table 3. The 700 # glass bonded grinding top is selected in the experiments with a diameter of 4 mm, while the abrasive grits are distributed over the 15 mm of tool length, grinding wheel specifications are listed in Table 4.

Table 2. Chemical compositions of K9 optical glass.

Element	SiO_2	B_2O_3	BaO	Na_2O	K_2O	As_2O_3
Content (wt%)	69.13	10.75	3.07	10.40	6.29	0.36

Table 3. Material properties of K9 glass used in the experiments.

Material	Dispersion	Elastic Modulus (GPa)	Fracture Toughness (MPa·m$^{1/2}$)	Vickers Hardness (GPa)	Poisson Ratio
K9	0.00806	88.5	2.63	7.8	0.203

Table 4. Grinding wheel specifications used in the experiments.

Type	Mesh	Wheel Diameter D_t	Grain Diameter D_d	Elastic Modulus	Poisson Ratio
PCD	700#	4 mm	21.8 μm	800 GP	0.07

The diameter size of the grinding rod is affected by the grinding parameters especially when the rotational spindle speed more than 50×10^4 min^{-1}, the diameter is equal or less than 1 mm, and the mesh size is almost less than 400#, all these conditions have adverse effects on the accuracy of grinding [33]. To avoid any possible deviation in the chip thickness calculations due to a decrease in the diameter of the grinding wheel, all these conditions will be taken into consideration in the current study.

5.2. Experimental Force

In the experimental part, Table 5 presents the grinding conditions. Kistler 9257 B dynamometer is used to gauge the grinding forces.

Table 5. Experimental conditions.

Group No	Feed Rate f_p (mm/min)	Spindle Speed N (min^{-1}) $\times 10^3$	Grinding Depth h (μm)	Grinding Width a_w (mm)	Tool Diameter D_t (mm)
1	60	10	10	1, 5, 10	4
2	150	5, 10, 15	10	5	4
3	60	10	5, 10, 15, 20	5	4
4	60, 100, 150, 200	5	10	5	4

The arithmetical model of the grinding force is established by assuming that the active grains on the grinding wheel surface are randomly distributed and the shape of all the grains is consistent. Besides, the workpiece materials selected in this study are considered as HBM, and the material properties will not be affected by the grinding conditions. The curves below presented in Figure 9 are the comparison between arithmetic values and measurement values of NT grinding forces under traditional surface-grinding conditions. It can be seen that the theoretical calculation results are consistent with the experimental outputs. The grinding target is to achieve high-quality surface and subsurface integrity, therefore the grinding force and grinding parameters must be under full control to achieve this goal. The precision of the prediction force represented in the mathematical model, grinding wheel conditions of the surface grinding process, fineness location of the workpiece, the ability to purify the measured grinding force through the output results are the essential reasons which create the deviations between the predicted force model and the experimental results.

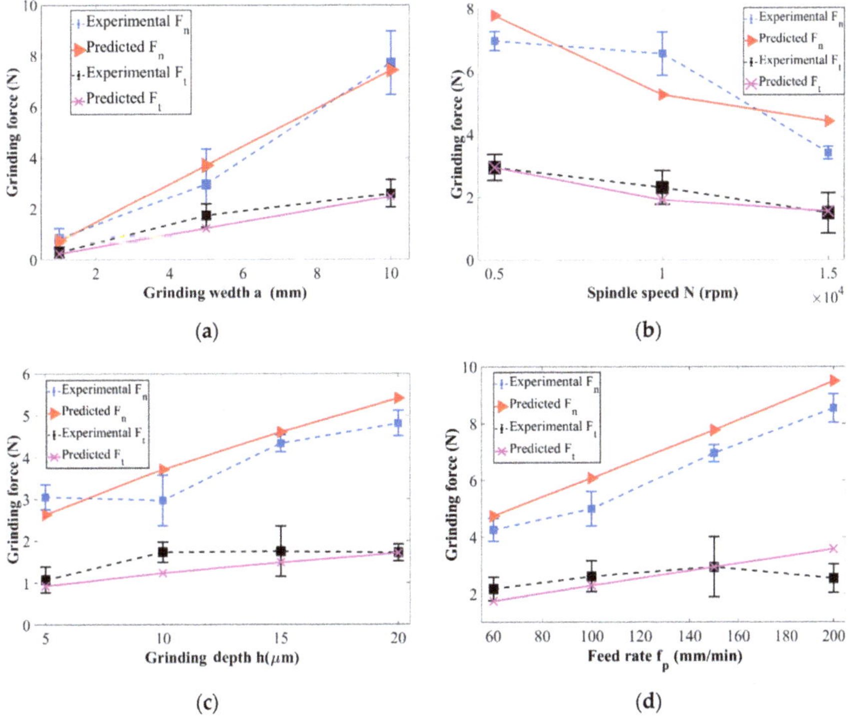

Figure 9. Experimental grinding forces at: (**a**) N = 10,000 min^{-1}, h = 10 μm, r_t = 2 mm, f_p = 60 mm/min; (**b**) a_w = 5 mm, h = 10 μm, r_t = 2 mm, f_p = 150 mm/min; (**c**) h = 10 μm, a_w = 5 mm, r_t = 2 mm, f_p = 60 mm/min; (**d**) N = 5000 min^{-1}, h = 10 μm, a_w = 5 mm, r_t = 2 mm.

5.3. Workpiece Topography

One of the most paramount demands in the industrial implementation is the quality of face topography, which can be counted as a leading indicator of production fineness. Figure 10 displays the workpiece surface captures taken by the ZEISS AXIO microscope apparatus. The captures are taken based on the grinding depth, the surface displayed in Figure 10a, which has a less grinding depth, seems to be better than Figure 10b–d. In conclusion, it can be clarified that brittle tear is the primary mode of elimination gradually with increasing chip thickness. Subsequently, the classifications of the material removal regions discussed in Section 2.2 are more agreeable.

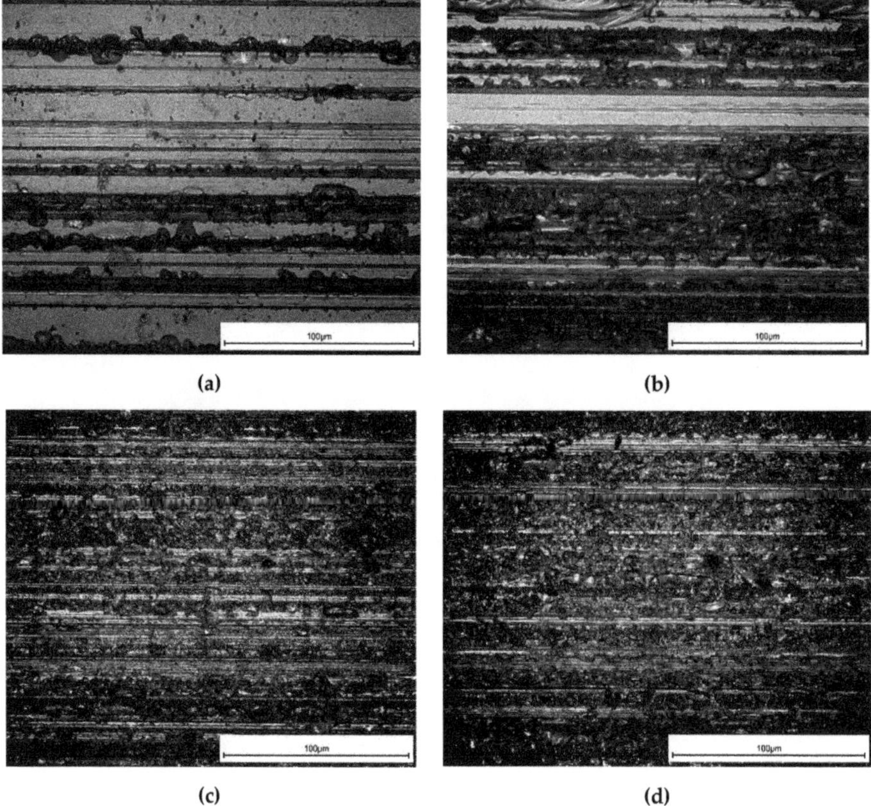

Figure 10. K9 surface topography at: (**a**) h = 1 μm, N = 10^4 min^{-1}, f_p = 60 mm/min; (**b**) h = 5 μm, N = 10^4 min^{-1}, f_p = 100 mm/min; (**c**) h = 10 μm, N = 10^4 min^{-1}, f_p = 60 mm/min; (**d**) h = 15 μm, N = 1.5×10^4 min^{-1}, f_p = 100 mm/min.

6. Conclusions

K9 glass has a high hardness and low fracture toughness which greatly fluctuates the cutting force generated during the machining process. Therefore, achieving the efficient prediction and controlling of the grinding force during the processing of optical glass is of great importance to improve the processing efficiency and the quality of the processed surfaces. The investigation model of the grinding force consists of ductile, ductile-to-brittle, brittle forces. The suggested force model is taking the two sources of friction into a consideration, as well as considers the elastic–plastic transition, randomly distributed of active grain, and the tool diameter variation during processing. Moreover, the predicted model is evaluated by comparing the experimental measurement results to the model prediction values. The main conclusions and results of this paper are as follows:

- The force model was verified by the K9 glass grinding test with a fixed abrasive grain through different grinding parameters. The theoretical analysis results have the same trend as the experimentally measured values, which proves that the model is reliable.
- Based on the analysis of the protruding height and horizontal distribution characteristics of abrasive grits on the grinding tool surface, the a_{max} is a dynamic concept, which has different values during grinding.
- Under the same grinding conditions, the grinding force inclines with the rotational speed N growing, where it decreases by growing the grinding depth h, grinding width a_w, and feed rate f_p. Besides, the conjunction impacts between the number of active grits and grit size led to the grinding force showing an irregular direction in the force model.
- The normal force is more than the tangential force during surface grinding of HBM, for that reason, controlling the forces in the normal axis is more significant.

Eventually, the current model is capable of fully representing the actual material removal attitudes and effectively predicting the NT forces when grinding the HBM surfaces as well.

Author Contributions: Conceptualization, H.M. and X.C.; Investigation, H.M., S.L., and X.G.; Software, H.M.; Validation, H.M. and X.C.; Writing, H.M. and G.Z.—original draft, H.M. and Y.L. All authors have read and agreed to the published version of the manuscript.

Funding: The paper is financially supported by the National Key Research and Development Project of China (2018 YFB2001400), and the Nature Science Foundation of Shandong Province (ZR2015 EL023).

Nomenclature

ϵ	Dimensionless constant related to HBM
μ	Coefficient of friction
a_c	Undeformed chip thickness
$a_{elas\text{-}plas}$, a_{cr}	Chip thickness of elastic-to-plastic, and ductile-to-brittle transitions
a_{max}	Maximum chip thickness
$A_{n\text{-}elas}$, $A_{t\text{-}elas}$	Normal and tangential areas in elastic zone respectively
$A_{n\text{-}plas}$, $A_{t\text{-}plas}$	Normal and tangential areas in plastic zone respectively
a_w	Grinding width
C_G	Size fracture of grinding grit.
E_p, E_d, E_{ef}	Workpiece, abrasive grain, and effective modulus of elasticity respectively
$F_{B.max}$	Maximum brittle force
F_n, F_t	Total normal and tangential grinding force
$F_{n.B}$, $F_{t.B}$	Normal and tangential brittle force
$F_{n.fr}$, $F_{t.fr}$	Normal and tangential frictional force
$F_{n\text{-}elas}$, $F_{t\text{-}elas}$	Normal and tangential force occurred on elastic zone
$F_{n\text{-}plas}$, $F_{t\text{-}plas}$	Normal and tangential force occurred on plastic zone
f_p	Feed rate
H_p	Workpiece hardness
h_t	Chip thickness in the elastic zone
K_{IC}	Fracture toughness of workpiece material
L_S, L_D, L_{ef}	Geometric, dynamic, and effective contact length between worksurface and grinding tool
N	Spindle speed
$N_{d\text{-}x}$, $N_{d\text{-}y}$, $N_{d\text{-}Tot}$	x-axis, y-axis, and total active grains share in grinding
P_0	Experiential suitability factor

P_{cont}, A_{cont}	Contact pressure and the contact area between the wear protrusion of grits and the grinding face
P_{max}	Maximum contact pressure
$P_{max\text{-}elas}$, $P_{max\text{-}plas}$	Maximum pressure created on worksurface through plastic and elastic regions respectively
R_c	Arc contact length between single grain and worksurface
r_d, D_d	Mean radius and diameter of grinding grain
r_t	Grinding wheel radius
$S(t)$	Displacement of a single grit
t	Grinding time
T_{ef}	Effective time
T_{Tot}	The total grinding time of grinding tool to intersect the dynamic touch length in one cycle
V_r	The relative velocity of a single grain relative to the workpiece surface
X_{max}	Maximum axial displacement of single abrasive grit in one cycle
α_1, α_2	Coefficients of the materiality of the frictional pairs
β	The half-oblique angle of a single active grit
ε, ε_c	Unitless constants
η, ζ	Constants defined according to the material sensitivity to strain average
λ_c	Touch arc length
λ_r	Average space between any two active grains
ϱ	Deviation of the arc of the grinding channel and the radius of the grinding rod
υ_p, υ_d	Poisson's ratios of workpiece and grit respectively
ω	Angular velocity

References

1. Zhao, J.; Huang, J.; Wang, R.; Peng, H.R.; Hang, W.; Ji, S. Investigation of the optimal parameters for the surface finish of K9 optical glass using a soft abrasive rotary flow polishing process. *J. Manuf. Process.* **2020**, *49*, 26–34. [CrossRef]
2. Liu, M.; Wu, J.; Gao, C. Sliding of a diamond sphere on K9 glass under progressive load. *J. Non-Cryst. Solids* **2019**, *526*, 119711. [CrossRef]
3. Guo, X.; Zhai, R.; Shi, Y.; Kang, R.; Jin, Z.; Guo, D. Study on influence of grinding depth and grain shape on grinding damage of K9 glass by SPH simulation. *Int. J. Adv. Manuf. Technol.* **2020**, *106*, 333–343. [CrossRef]
4. Zhao, P.Y.; Zhou, M.; Liu, X.L.; Jiang, B. Effect to the surface composition in ultrasonic vibration-assisted grinding of BK7 optical glass. *Appl. Sci.* **2020**, *10*, 516. [CrossRef]
5. Li, J.; Wang, H.; Wang, W.; Huang, J.; Zhu, Y.; Zuo, D. Model of surface roughness in fixed abrasive lapping of K9 glass. *J. Mech. Eng.* **2015**, *51*, 199–205. [CrossRef]
6. Cheng, J.; Gong, Y.; Yan, X.; Zheng, W. Modeling and experimental study of complex critical condition for ductile-regime micro-grinding of hard brittle material. *J. Mech. Eng.* **2013**, *49*, 191–198. [CrossRef]
7. Li, P.; Jin, T.; Xiao, H.; Chen, Z.; Qu, M.; Dai, H.; Chen, S. Topographical characterization and wear behavior of diamond wheel at different processing stages in grinding of N-BK7 optical glass. *Tribol. Int.* **2020**, *151*, 106453. [CrossRef]
8. Blaineau, P.; André, D.; Laheurte, R.; Darnis, P.; Darbois, N.; Cahuc, O.; Neauport, J. Subsurface mechanical damage during bound abrasive grinding of fused silica glass. *Appl. Surf. Sci.* **2015**, *353*, 764–773. [CrossRef]
9. Sun, Y.; Su, Z.P.; Gong, Y.D.; Jin, L.Y.; Wen, Q.; Qi, Y. An experimental and numerical study of micro-grinding force and performance of sapphire using novel structured micro abrasive tool. *Int. J. Mech. Sci.* **2020**, *181*, 105741. [CrossRef]
10. Yang, Z.; Zhu, L.; Lin, B.; Zhang, G.; Ni, C.; Sui, T. The grinding force modeling and experimental study of ZrO 2 ceramic materials in ultrasonic vibration assisted grinding. *Ceram. Int.* **2019**, *45*, 8873–8889. [CrossRef]
11. Ni, C.; Zhu, L.; Liu, C.; Yang, Z. Analytical modeling of tool-workpiece contact rate and experimental study in ultrasonic vibration-assisted milling of Ti–6Al–4V. *Int. J. Mech. Sci.* **2018**, *142*, 97–111. [CrossRef]
12. Zhang, X.; Li, C.; Zhang, Y.; Wang, Y.; Li, B.; Yang, M.; Guo, S.; Liu, G.; Zhang, N. Lubricating property of MQL grinding of Al_2O_3/SiC mixed nanofluid with different particle sizes and microtopography analysis by cross-correlation. *Precis. Eng.* **2017**, *47*, 532–545. [CrossRef]

13. Zhu, L.; Yang, Z.; Li, Z. Investigation of mechanics and machinability of titanium alloy thin-walled parts by CBN grinding head. *Int. J. Adv. Manuf. Technol.* **2019**, *100*, 2537–2555. [CrossRef]
14. Liu, S.; Chen, T.; Wu, C. Rotary ultrasonic face grinding of carbon fiber reinforced plastic (CFRP): A study on cutting force model. *Int. J. Adv. Manuf. Technol.* **2017**, *89*, 847–856. [CrossRef]
15. Zhang, C.; Zhang, J.; Feng, P. Mathematical model for cutting force in rotary ultrasonic face milling of brittle materials. *Int. J. Adv. Manuf. Technol.* **2013**, *69*, 161–170. [CrossRef]
16. Sun, G.; Zhao, L.; Ma, Z.; Zhao, Q. Force prediction model considering material removal mechanism for axial ultrasonic vibration-assisted peripheral grinding of Zerodur. *Int. J. Adv. Manuf. Technol.* **2018**, *98*, 2775–2789. [CrossRef]
17. Jianhua, Z.; Hui, L.; Minglu, Z.; Yan, Z.; Liying, W. Study on force modeling considering size effect in ultrasonic-assisted micro-end grinding of silica glass and Al_2O_3 ceramic. *Int. J. Adv. Manuf. Technol.* **2017**, *89*, 1173–1192. [CrossRef]
18. Xiao, X.; Zheng, K.; Liao, W.; Meng, H. Study on cutting force model in ultrasonic vibration assisted side grinding of zirconia ceramics. *Int. J. Mach. Tools Manuf.* **2016**, *104*, 58–67. [CrossRef]
19. Badger, J.A.; Torrance, A.A. Comparison of two models to predict grinding forces from wheel surface topography. *Int. J. Mach. Tools Manuf.* **2000**, *40*, 1099–1120. [CrossRef]
20. Chen, S.T.; Jiang, Z.H. A force controlled grinding-milling technique for quartz-glass micromachining. *J. Mater. Process. Technol.* **2015**, *216*, 206–215. [CrossRef]
21. Su, Y.; Lin, B.; Cao, Z. Prediction and verification analysis of grinding force in the single grain grinding process of fused silica glass. *Int. J. Adv. Manuf. Technol.* **2018**, *96*, 597–606. [CrossRef]
22. Jian-Hua, Z.; Yan, Z.; Fu-Qiang, T.; Shuo, Z.; Lan-Shen, G. Kinematics and experimental study on ultrasonic vibration-assisted micro end grinding of silica glass. *Int. J. Adv. Manuf. Technol.* **2015**, *78*, 1893–1904. [CrossRef]
23. Li, Z.; Zhang, F.; Luo, X.; Guo, X.; Cai, Y.; Chang, W.; Sun, J. A new grinding force model for micro grinding RB-SiC ceramic with grinding wheel topography as an input. *Micromachines* **2018**, *9*, 368. [CrossRef]
24. Lee, S.H. Analysis of ductile mode and brittle transition of AFM nanomachining of silicon. *Int. J. Mach. Tools Manuf.* **2012**, *61*, 71–79. [CrossRef]
25. Yang, M.; Li, C.; Zhang, Y.; Jia, D.; Zhang, X.; Hou, Y.; Li, R.; Wang, J. Maximum undeformed equivalent chip thickness for ductile-brittle transition of zirconia ceramics under different lubrication conditions. *Int. J. Mach. Tools Manuf.* **2017**, *122*, 55–65. [CrossRef]
26. Cheng, J.; Wu, J.; Gong, Y. Ductile to brittle transition in ultra-micro-grinding (UMG) of hard brittle crystal material. *Int. J. Adv. Manuf. Technol.* **2018**, *97*, 1971–1994. [CrossRef]
27. Kar, S.; Kumar, S.; Bandyopadhyay, P.P.; Paul, S. Grinding of hard and brittle ceramic coatings: Force analysis. *J. Eur. Ceram. Soc.* **2020**, *40*, 1453–1461. [CrossRef]
28. Wang, Z.; Li, H.N.; Yu, T.B.; Chen, H.; Zhao, J. On the predictive modelling of machined surface topography in abrasive air jet polishing of quartz glass. *Int. J. Mech. Sci.* **2019**, *152*, 1–18. [CrossRef]
29. Pratap, A.; Patra, K. Combined effects of tool surface texturing, cutting parameters and minimum quantity lubrication (MQL) pressure on micro-grinding of BK7 glass. *J. Manuf. Process.* **2020**, *54*, 374–392. [CrossRef]
30. Zhang, X.; Kang, Z.; Li, S.; Shi, Z.; Wen, D.; Jiang, J.; Zhang, Z. Grinding force modelling for ductile-brittle transition in laser macro-micro-structured grinding of zirconia ceramics. *Ceram. Int.* **2019**, *45*, 18487–18500. [CrossRef]
31. Zhang, Y.; Li, C.; Ji, H.; Yang, X.; Yang, M.; Jia, D.; Zhang, X.; Li, R.; Wang, J. Analysis of grinding mechanics and improved predictive force model based on material-removal and plastic-stacking mechanisms. *Int. J. Mach. Tools Manuf.* **2017**, *122*, 81–97. [CrossRef]
32. Cheng, J.; Gong, Y.D. Experimental study of surface generation and force modeling in micro-grinding of single crystal silicon considering crystallographic effects. *Int. J. Mach. Tools Manuf.* **2014**, *77*, 1–15. [CrossRef]
33. Li, W.; Ren, Y.; Li, C.; Li, Z.; Li, M. Investigation of machining and wear performance of various diamond micro-grinding tools. *Int. J. Adv. Manuf. Technol.* **2020**, *106*, 921–935. [CrossRef]
34. Li, C.; Li, X.; Wu, Y.; Zhang, F.; Huang, H. Deformation mechanism and force modelling of the grinding of YAG single crystals. *Int. J. Mach. Tools Manuf.* **2019**, *143*, 23–37. [CrossRef]
35. Malkin, S.; Guo, C. *Grinding Technology Theory & Application of Machining with Abrasives*, 2nd ed.; Industrial Press: New York, NY, USA, 2008; Volume 2, ISBN 9780831132477.

Micromachines **2020**, *11*, 969

36. Pelletier, H.; Gauthier, C.; Schirrer, R. Friction effect on contact pressure during indentation and scratch into amorphous polymers. *Mater. Lett.* **2009**, *63*, 1671–1673. [CrossRef]

37. Sadhal, S.S. *Contact Mechanics*; Springer: Berlin/Heidelberg, Germnay, 2011; Volume 279. [CrossRef]

38. Popov, V.L. *Contact Mechanics and Friction—Hertz Force*; Springer: Berlin/Heidelberg, Germany, 2017; ISBN 978-3-642-10802-0.

39. Jamshidi, H.; Budak, E. An analytical grinding force model based on individual grit interaction. *J. Mater. Process. Technol.* **2020**, *283*, 116700. [CrossRef]

40. Zhang, W.; Subhash, G. An elastic-plastic-cracking model for finite element analysis of indentation cracking in brittle materials. *Int. J. Solids Struct.* **2001**, *38*, 5893–5913. [CrossRef]

41. El-Taybany, Y.; El-Hofy, H. Mathematical model for cutting force in ultrasonic-assisted milling of soda-lime glass. *Int. J. Adv. Manuf. Technol.* **2019**, *103*, 3953–3968. [CrossRef]

42. Li, B.; Dai, C.; Ding, W.; Yang, C.; Li, C.; Kulik, O.; Shumyacher, V. Prediction on grinding force during grinding powder metallurgy nickel-based superalloy FGH96 with electroplated CBN abrasive wheel. *Chin. J. Aeronaut.* **2020**. [CrossRef]

43. Zhou, M.; Zheng, W. A model for grinding forces prediction in ultrasonic vibration assisted grinding of SiCp/Al composites. *Int. J. Adv. Manuf. Technol.* **2016**, *87*, 3211–3224. [CrossRef]

44. Lu, S.; Gao, H.; Bao, Y.; Xu, Q. A model for force prediction in grinding holes of SiCp/Al composites. *Int. J. Mech. Sci.* **2019**, *160*, 1–14. [CrossRef]

45. Li, C.; Zhang, F.; Ding, Y.; Liu, L. Surface deformation and friction characteristic of nano scratch at ductile-removal regime for optical glass BK7. *Appl. Opt.* **2016**, *55*, 6547. [CrossRef] [PubMed]

46. Huang, H.; Yin, L. Grinding characteristics of engineering ceramics in high speed regime. *Int. J. Abras. Technol.* **2007**, *1*, 78–93. [CrossRef]

Publisher's Note: MDPI stays neutral with regard to jurisdictional claims in published maps and institutional affiliations.

 *micromachines*

Article

Molecular Dynamics Study of the Effect of Abrasive Grains Orientation and Spacing during Nanogrinding

Nikolaos E. Karkalos and Angelos P. Markopoulos * 

Laboratory of Manufacturing Technology, School of Mechanical Engineering, National Technical University of Athens, Heroon Polytechniou 9, 15780 Athens, Greece; nkark@mail.ntua.gr
* Correspondence: amark@mail.ntua.gr; Tel.: +30-210-772-4299

Received: 20 June 2020; Accepted: 21 July 2020; Published: 23 July 2020

Abstract: Grinding at the nanometric level can be efficiently employed for the creation of surfaces with ultrahigh precision by removing a few atomic layers from the substrate. However, since measurements at this level are rather difficult, numerical investigation can be conducted in order to reveal the mechanisms of material removal during nanogrinding. In the present study, a Molecular Dynamics model with multiple abrasive grains is developed in order to determine the effect of spacing between the adjacent rows of abrasive grains and the effect of the rake angle of the abrasive grains on the grinding forces and temperatures, ground surface, and chip formation and also, subsurface damage of the substrate. Findings indicate that nanogrinding with abrasive grains situated in adjacent rows with spacing of 1 Å leads directly to a flat surface and the amount of material remaining between the rows of grains remains minimal for spacing values up to 5 Å. Moreover, higher negative rake angle of the grains leads to higher grinding forces and friction coefficient values over 1.0 for angles larger than −40°. At the same time, chip formation is suppressed and plastic deformation increases with larger negative rake angles, due to higher compressive action of the abrasive grains.

Keywords: nanogrinding; abrasive grains; rake angle; spacing; grinding forces; grinding temperature; chip formation; subsurface damage

1. Introduction

Ultrahigh precision machining plays an increasingly important role in various contemporary high-end industries due to the necessity of fabricating parts or rendering features on various components with micrometer or nanometer dimensions. The strict requirements for ensuring dimensional accuracy of mechanical parts in an extremely small scale render the procedure of selecting the appropriate manufacturing technique considerably important [1]. Some of the most widely employed techniques for manufacturing in the nanoscale level are related to nanolithography technique, whereas in areas such as the production of high quality and precise lens, ultrahigh precision polishing has been proven adequate for achieving roughness value as low as several nanometers [1,2].

Although the aforementioned techniques, such as nanometric level polishing have been already employed in industrial scale, material removal mechanisms at nanometric level have not been yet studied in every aspect. Moreover, the inability of conducting appropriate experiments due to high cost or lack of adequate equipment leads to the necessity of employing alternative methods for the study of nanoscale machining and abrasive processes, by conducting dedicated and reliable simulations [3]. For the macroscale and microscale counterparts of these processes, the finite element method (FEM) has been successfully employed for decades, as it can achieve a sufficient level of accuracy and has been assistive to the understanding of various underlying phenomena. Despite the fact that by using special models, microscale phenomena can be modeled with FEM, the discrete nature of matter in the nanoscale level and inability of continuum theories to represent atomistic interactions led to the use of Molecular Dynamics (MD) method as a viable alternative for nanoscale simulations.

The MD method, developed originally in the 1950s by Alder and Wainwright [4,5], is an established computational method for studying various phenomena and processes related to the nanoscale, such as the determination of fundamental material structure and properties. This method models directly the atomistic structure of materials and approximates the material behavior in the nanoscale level with a high degree of accuracy. Although in most cases, a direct comparison with experimental data is not feasible, MD method can provide adequate explanations regarding phenomena which have their origins in the nanoscale, such as plastic deformation and material removal mechanism during machining and abrasive process, and thus, can act complementary to macroscale experiments and simulations by supporting their findings. In the field of nanocutting simulations, it has been applied during the last three decades, stemming from the pioneering works of Belak and Stowers [6] and Belak, Boercker, and Stowers [7]. These early studies, although they were conducted with relatively small and simple models, set the path for the more advanced models which were presented during the last three decades. It is worth noting that although most models were simpler than the current ones, the authors distinguished the nanocutting from the nanoabrasive models, which included one or two abrasive grains with a linear motion. During the last decades, considerable progress has been conducted in this field, especially regarding nanogrinding simulations, which will be briefly discussed afterwards.

Inasaki and Rentsch [8,9] were among the first to investigate the mechanisms of material removal, as well as surface integrity during nanogrinding. In one of their earliest studies [8], they conducted simulations with a single abrasive grain at two different orientations and discussed various important aspects of the nanometric abrasive simulation. Moreover in a later work [9], they simulated both ductile and brittle materials and studied not only the material removal mechanism and dislocation evolution, but also crack propagation for a silicon substrate. The effect of abrasive grain rake angle in nanometric grinding was studied by Komanduri et al. [10] with the aid of an appropriate MD model. More specifically in this work, the effect of rake angle on cutting forces and energy for a wide range of rake angle values from +45° to −75° was investigated with a rigid diamond tool. It was found that both cutting and thrust force increased for decreasing rake angle, with thrust force being mostly influenced as it increased almost 17 times, whereas a threefold increase was observed for cutting force; furthermore, the ratio of force components exceeded 1.0, with a value of 2.361 at a rake angle of −75°. Lin et al. [11] presented an MD nanogrinding model with a single hemispherical abrasive grain. Their study focused on material removal mechanisms and surface alterations of a silicon workpiece. With this model, they were able to observe chip formation and plastic deformation on the substrate and their analysis showed the correlation between grinding force fluctuations and dislocation movement.

In the relevant literature, there are also several computational studies with the use of MD method, related to high-speed grinding. Li et al. [12] investigated the effect of various process parameters such as grinding speed, depth of cut, and abrasive grain radius on chip formation and subsurface damage in high-speed grinding. Shimizu et al. [13] investigated the effect of grinding speed in relation to the wave propagation speed of the workpiece material with an MD model with a single hemispherical grain. It was found that, at lower speeds, increase of grinding speed resulted in lower plastic deformation but when the wave propagation speed was exceeded, severe hardening of the workpiece occurred. Li et al. [14] used an MD model to study the effect of grinding speed in nanoscale grinding of silicon and were able to determine the ductile to brittle cutting mode transition and the relation of stresses and subsurface damage. Guo et al. [15] presented an MD study of nanogrinding considering the effect of multiple passes of the abrasive grain on the workpiece. In the simulation, four passes of the abrasive grain were performed and after the determination of subsurface damage, it was concluded that the optimum depth of cut should not be higher than the half of the initial depth of cut.

Chen et al. [16] investigated the residual stresses during nanogrinding of silicon with a cylindrical tool. Simulations were carried out for various grinding depths from 5–30 Å. It was found that as the depth increased, although the region of transition between compressive and tensile residual stress remained almost the same, an increase of the compressive stresses near the surface was observed. Ren et al. [17] performed simulations regarding nanogrinding of monocrystalline nickel under various process conditions. Based on their findings, they observed an increase of cutting forces, potential, and temperature values with increased depth of cut and grinding speed values. Liu et al. [18] studied the effect of grinding speed and grinding depth on forces, stresses, temperature, and specific energy during nanogrinding of silicon carbide (SiC) with a conical tool. The increase of grinding speed and depth affected all quantities positively, with the greatest increase observed in the case of tangential grinding force and temperature. Liang et al. [19] presented a nanogrinding model with a cylindrical roller as cutting tool and investigated the influence of rotating speed and workpiece temperature on grinding forces. Higher rotational speed and increased initial workpiece temperature resulted in higher grinding forces.

Zhang et al. [20] investigated the surface damage layer formation during nanogrinding and its effect on workpiece mechanical properties. At first, they determined the optimum grinding speed for obtaining minimum thickness of damage layer and then performed simulations of tensile tests on the deformed workpieces. It was found that although Young's modulus remained almost unaffected by the damage induced by nanogrinding, UTS was considerably affected, up to 47%. Xu et al. [21] also conducted a study regarding material removal and amorphous layer formation during nanometric grinding of a silicon wafer containing SiO_2 region. They showed that, increase of grinding speed was able to reduce the damage layer depth and cutting forces, whereas friction coefficient increased with a large grinding depth. Ren et al. [22] performed nanogrinding simulations of monocrystalline nickel with a spherical abrasive grain and determined the effect of process parameters such as grinding speed, depth of cut, and abrasive grain radius on subsurface damage. At lower grinding speeds, stacking faults were formed in the substrate which were reduced in size gradually at higher grinding speeds and eventually vanished at 400 m/s. The increase of grinding depth and abrasive grain diameter led to an increase of stacking faults and dislocations, whereas the size of the subsurface damage layer was found to be almost unaffected by the grinding speed. The same scientific group, in a later study [23] investigated the effect of crystal orientation in the case of nanocrystalline nickel. After they carried out simulations for six different crystal orientations, they showed that burr height was minimum at $(110)[\bar{1}10]$ direction, whereas the deformed layer depth was minimum at $(111)[\bar{1}10]$ direction. Liu et al. [24] studied the effect of grinding speed and depth of cut on a SiC substrate. Their findings indicated that higher grinding speed leads to higher temperature but lower forces and lower thickness of amorphous layer, whereas higher depth of cut leads to higher temperature, higher forces, and higher thickness of amorphous layer.

Apart from silicon and common metals such as Cu, Al, and Ni, recently several studies focused also on the nanogrinding of bilayer workpieces. More specifically, Xu et al. [25] performed nanogrinding simulations for Cu-Si and $Cu-SiO_2$ bilayer workpieces with a hemispherical tool. In their simulations, they employed various grinding depth and speed values and analyzed the deformation of the workpiece. Wang et al. [26] conducted a comprehensive study regarding nanogrinding of Al-Si bilayers under various grinding depth, speed, and abrasive grain radius values. They showed that an increase of the grain radius led to increase of workpiece temperature and grinding forces, and the same conclusions were drawn regarding grinding depth and speed. Moreover, the Si-Al bilayers exhibited higher forces and temperatures than the Al-Si bilayers. Fang et al. [27] investigated the material removal and surface integrity of Ni/Cu bilayers. Similarly to the study of Wang et al. [26], they found also that the increase of grinding speed, depth, and abrasive grain radius lead to increase of both grinding forces and temperature of the workpiece but it was also observed that the highest temperatures occurred in Ni/Cu bilayers, whereas the largest forces on simple Ni workpieces. Furthermore, lower abrasive grain radius and grinding depth led to better quality and lower damage. Finally, Xu et al. [28] also presented a thorough study on nanogrinding of Cu-Si substrates and examined the influence of copper layer thickness as well as grinding depth and speed on forces, temperature, potential and kinetic energy, and dislocation length, showing their correlations and their impact on phase transformation of the substrate as well.

In the previous studies, most models included a single abrasive grain at a linear motion. However, some researchers employed more sophisticated models regarding abrasive trajectory and shape. For example, Fang et al. [29] presented a model in which the abrasive grain moved at a complex trajectory, containing both a straight and an arc-shaped path. Eder et al. [30,31] developed a model for nanopolishing using cubic and spherical abrasive grains with random orientations while the workpiece surface was modeled using a pseudorandom Gaussian topography. Nevertheless, there is still a lack for models for peripheral nanogrinding using multiple abrasive grains.

Thus, in the present work, an MD model of peripheral nanogrinding is presented using multiple abrasive grains. The trajectory of the grains is consistent with the kinematics of peripheral nanogrinding and grains have different protrusion height as well. With the aid of this model, two sets of simulations are conducted; the first set of simulations is conducted in order to determine the effect of spacing between the two rows of abrasive grains on ground surface and chip formation, whereas the second set of simulations focuses on the effect of the rake angle of the abrasive grains on grinding forces, chip formation, and subsurface damage of the workpiece. In both cases, the results regarding the aforementioned quantities are discussed and conclusions on the effect of spacing on surface and chip formation as well as the effect of the negative rake angle on the outcome of the process are drawn.

2. Materials and Methods

In the present study, an MD model regarding peripheral nanogrinding with multiple abrasive grains is presented. The aim of this study was the determination of the effect of spacing of the two rows of abrasive grains on surface and chip formation, as well as the effect of abrasive grain angle on grinding forces, friction coefficient, and subsurface damage. Thus, two different sets of simulations were carried out, with the necessary adjustments performed on the MD model. A general view of the model is depicted in Figure 1.

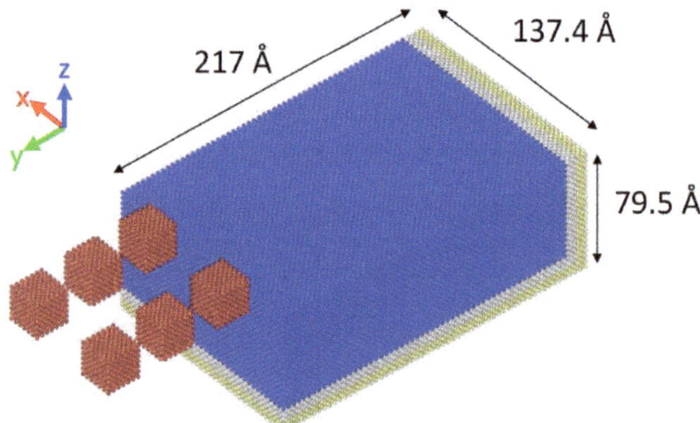

Figure 1. Schematic of the Molecular Dynamics (MD) model.

In the first set of simulations, 6 different spacing values were employed. For convenience, these values were defined in terms of the lattice parameter of tool material, denoted as a (a = 3.57 Å): 1a, 3a, 5a, 7a, 9a, and 11a. This wide range of values allowed for studying of considerably different cases in order to be able to determine more clearly the various states of the ground surface in respect to the process conditions. In each case, grinding length was 160 Å. The respective grains of each adjacent row had the same protrusion height and different position in the x axis, depending on the spacing in each case. The two rows of abrasive grains were situated in symmetrical positions in respect to the axis of symmetry of the workpiece in the x axis. The first set of abrasive grains, which were closest to the workpiece, was positioned initially at a depth of 2a and each following set was positioned 0.5a below the previous set of abrasive grains. In every case, the rake angle of the abrasive grains was 0°. For the second set of simulations, the same model was employed but the spacing was fixed at 3a, whereas the rake angle of the abrasive grains varied in the range of −5° to −60°. At each simulation, the rake angle of all abrasive grains was the same.

Workpiece material was monocrystalline copper, whereas the abrasive grains were diamond single crystal grains. The workpiece material was modeled with embedded atom model (EAM) potential function [32], which was particularly capable of describing the interactions due to metallic bonds as it takes into consideration the effects of electron density. For the interaction between abrasive grains and workpiece, Morse potential function for Cu-C pairs was employed [33]. The number of atoms for the MD model was 210,000. Furthermore, the dimensions of the copper workpiece along the three axes (xyz) was 217 Å × 137.4 Å × 79.5 Å.

As can be seen in Figure 1, the workpiece includes three different regions of atoms, namely, the region of Newtonian atoms, situated at the upper part of the workpiece and depicted in blue color, the region of thermostat atoms, situated below the Newtonian atoms region and depicted in white color, and the region of boundary atoms, situated at the right and bottom faces of the workpiece and depicted in beige color. The region of Newtonian atoms represents the main bulk of the workpiece, which is assumed to be deformable and atoms are allowed to move according to forces produced by their interaction with other atoms of the substrate or the atoms of the abrasive grains. The region of thermostat atoms is related to the regulation of workpiece temperature and dissipation of excessive heat by rescaling the velocity of these atoms [22] and finally, the region of fixed boundary atoms is essential to avoid rigid body motion and reduce the boundary effects. Moreover, periodic boundary conditions are imposed on the workpiece boundaries on the YZ planes. The diamond abrasive grains are considered much harder than the workpiece material and thus, are assumed to be a rigid body. For that reason, no interaction between C-C atoms is defined. The atoms of the workpiece are initially assigned velocities according to Maxwell-Boltzmann distribution based on the desired initial temperature and after a sufficient period for thermalization of the workpiece, nanogrinding takes place. The environment of the nanogrinding system is assumed to be vacuum, thus effects of grinding fluid or slurry are not taken into consideration.

The abrasive grains follow a complex path, including the influence of the rotation of the grinding wheel, which has a surface speed of 108 m/s and also incorporating the feed of the workpiece (v_f), which is equal to 100 m/s, as the workpiece is fixed in the simulations. Feed direction is towards the −y axis, whereas rotation is taking place in the YZ plane. This complex movement is represented by a type of trochoid curve with the motion of the center or mass of each abrasive particle being described by Equations (1) and (2):

$$x(t) = x_0 + R\cos(\omega t + \varphi_0) + v_f t \tag{1}$$

$$y(t) = y_0 + R\sin(\omega t + \varphi_0) \tag{2}$$

In these equations, x_0 and y_0 represent the coordinates of the initial point where the center of mass of the abrasive grains was situated, R is the radius of the grinding wheel on which the abrasive grains are supposed to be placed, and ω is the radial velocity of the grinding wheel. The trajectory was determined in such a way that each abrasive grain loses contact with the workpiece only when it is close to the final grinding length. Initial temperature of the workpiece was set to 293 K and numerical time-step for the simulations was 1 fs. All simulations were carried out using LAMMPS software.

3. Results and Discussion

3.1. Effect of Abrasive Grains Spacing on Surface and Chip Formation and Subsurface Damage

After the simulations were carried out, results regarding the first set of simulations will be discussed. The effect of spacing between the two rows of abrasive grains on the surface formed after the action of the abrasive grains and the produced chip can be observed in the snapshots of Figure 2. In this figure, snapshots of the last timestep of each case are being depicted for cases starting for spacing values from 1a to 11a. Moreover, in Figure 3, a top view of the workpiece at the same timesteps is presented with the atoms colored according to their z coordinate. From these results it can be considered that there are significant differences between these cases, something that will be examined in detail.

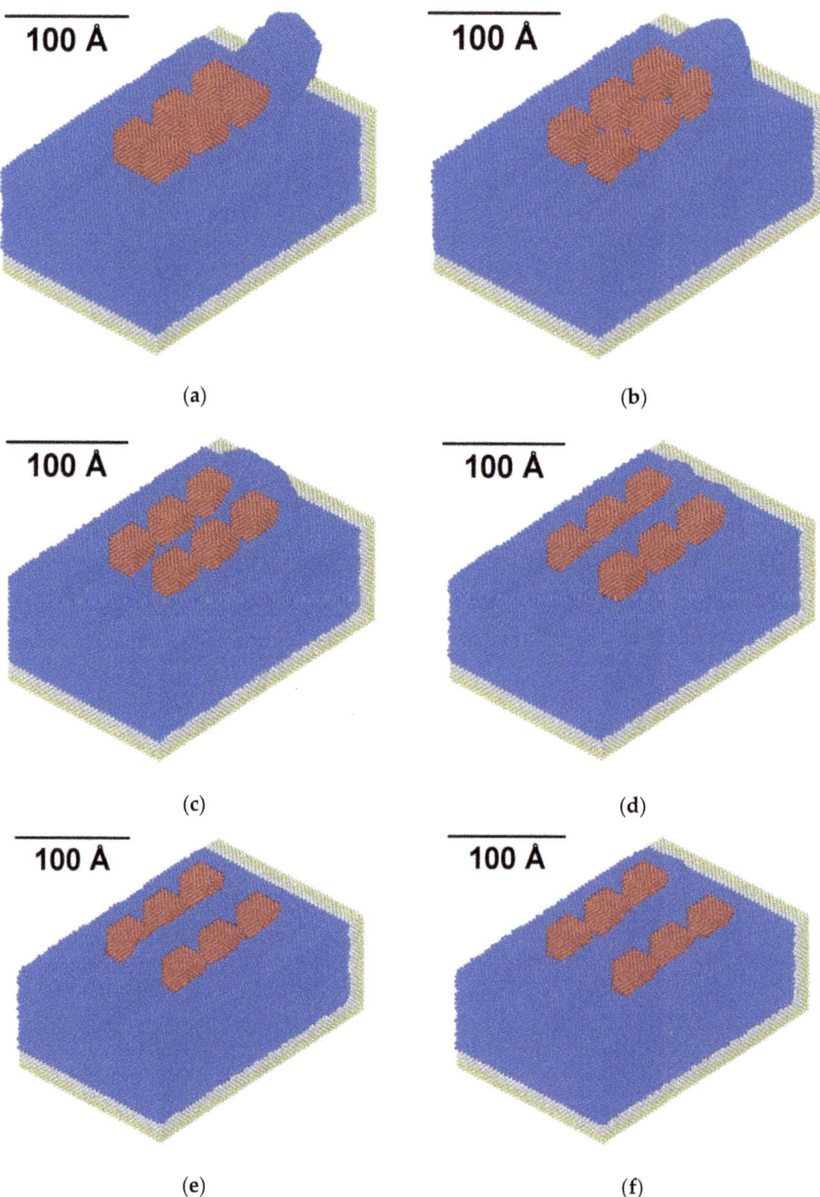

Figure 2. Snapshots of nanogrinding at the final timestep for the cases with spacing values of: (**a**) 1a, (**b**) 3a, (**c**) 5a, (**d**) 7a, (**e**) 9a, and (**f**) 11a.

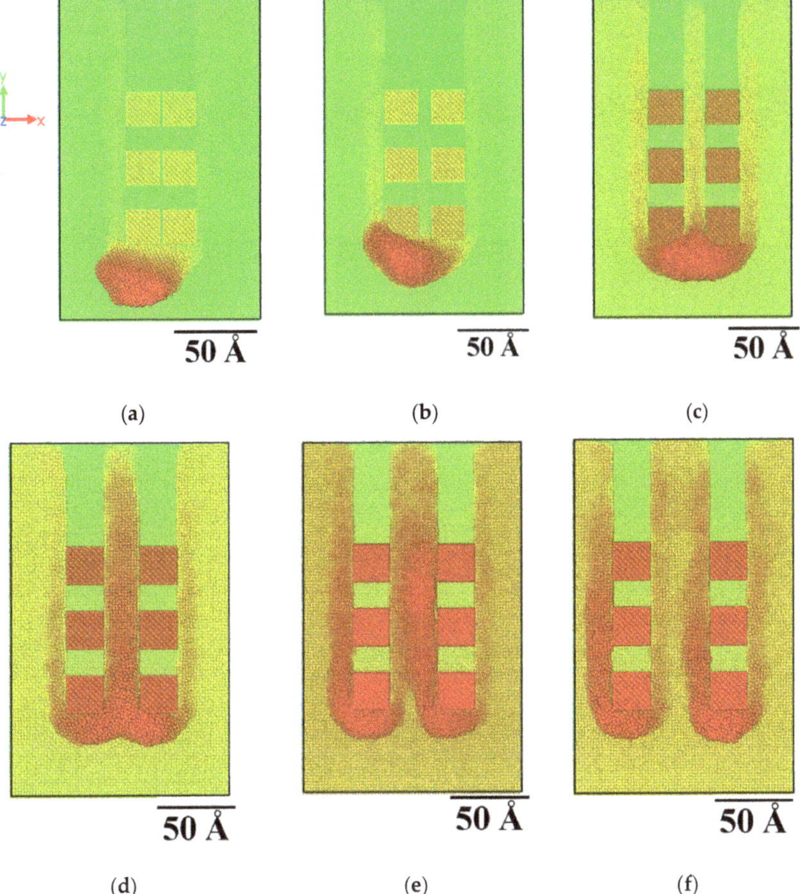

Figure 3. Top view of the workpiece at the final timestep for the cases with a spacing of: (**a**) 1a, (**b**) 3a, (**c**) 5a, (**d**) 7a, (**e**) 9a, and (**f**) 11a.

At first, it can be observed, from both Figures 2 and 3, that due to the very close distance between the abrasive grains, a single groove is formed from the action of the two rows of abrasive grains, as their interaction with the workpiece material removes all atoms along their path, including atoms situated between them. As the spacing increases, gradually a zone of atoms is formed between the two rows, and for spacing values from 5a and beyond, two separate grooves exist. However, the chip produced by the material removal from both rows of abrasive grains is still common until spacing value exceeds 7a. In these cases, the chip produced by each row of abrasive grains is separate. Apart from the chip in the front of the abrasive grains, there is also an amount of atoms deposited on the area between the two grooves and on the sides of the grooves, especially in the cases with larger spacing.

In order to quantify the effect of spacing on chip dimension, the chip height was also measured in each case, as can be seen in Figure 4. It is to be noted that the chip height is calculated from the top of the initial workpiece surface. The chip height is shown to become constantly lower as spacing increases before eventually stabilizing around 100 Å. From these observations, it can be concluded that the largest reduction of chip height actually coincides with the beginning of the formation of two separate grooves and then, stabilization of maximum chip height occurs after the chips produced by the two adjacent rows of abrasive grains are separated.

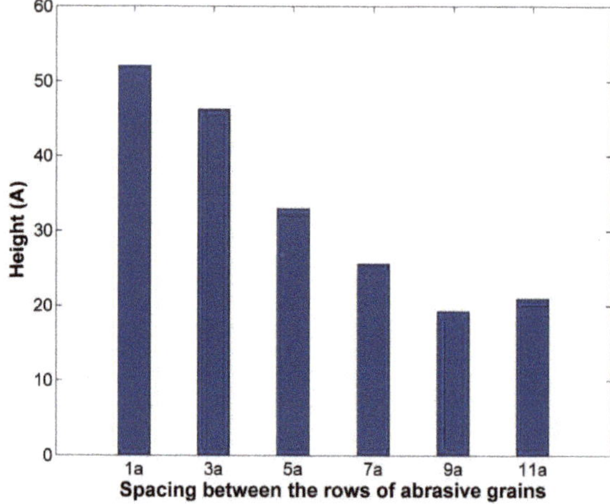

Figure 4. Height of chip in cases with different spacing between the rows of abrasive grains.

The observations from the snapshots of nanogrinding and chip height measurements can be further supported by the computed values of average temperature in the region of Newtonian atoms. As the spacing becomes larger, it can be seen from Figure 5 that average temperature values are gradually rising from a temperature of 472.65 K for a spacing of 1a and eventually, they stabilize at 530–540 K. Temperature is directly connected with variations in the kinetic energy of the atoms and thus, an increase of temperature indicates either that the atoms which are affected by the grinding action have increased velocities or that more atoms are affected by the action of the abrasive grains. As the process parameters are the same but the larger spacing between the grains leads to a wider affected area, the second case is more probable. What is more interesting, the most abrupt changes in the average temperatures coincide with the cases which mark the transitions from the formation of a single groove to the formation of separate grooves and finally, separate pileups of atoms, i.e., after the cases with a spacing of 3a and 7a, respectively.

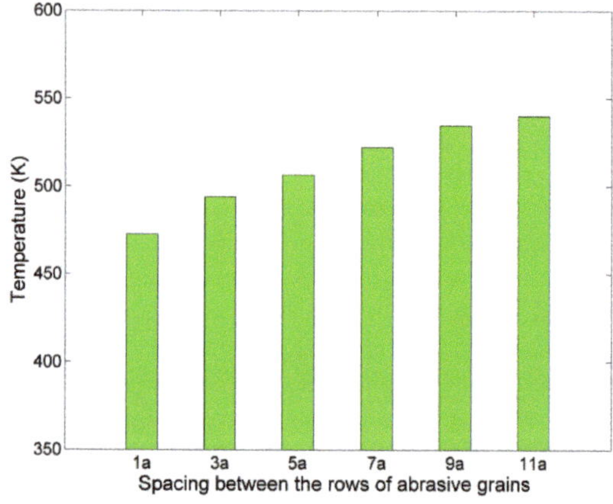

Figure 5. Average temperature of Newtonian atoms in cases with different spacing between the rows of abrasive grains.

The results of this set of simulations can be useful for the determination of optimum grinding strategy towards the rendering of a high-quality flat surface on the workpiece. For example, when the spacing between the adjacent rows of abrasive grains is larger, a large amount of material remains between the rows of grains and different process conditions and number of passes in x axis are required to render a flat surface, than in a case where almost no material is left between the grains. On the other hand, narrower spacing between the grains leads to material removal in a smaller area. Thus, in fact in order to determine the strategy required to achieve the highest efficiency, a detailed study based on the findings of the present work should be conducted.

3.2. Effect of Abrasive Grain Rake Angle on Grinding Forces, Chip Formation, and Subsurface Damage

After the results regarding the first set of simulations were presented and discussed, the analysis of the results of the second set of simulations is taking place. In this set of simulations, the effect of abrasive grain rake angle is investigated for negative values starting from $-5°$ to $-60°$. Although no similar works regarding the rake angle of abrasive grains are reported, this range of values was selected in accordance with the relevant literature on nanocutting simulations with negative rake angle tools [10,34–48]. More specifically, most authors have chosen negative rake angle values up to $-45°$ [34,43,45] or $-60°$ [10,35,38], and Lai et al. [37] pointed out that the critical rake angle for formation of chip is $-65°$, as they proved that higher negative rake angles resulted in plastic deformation of the substrate without a pileup of atoms.

In Figure 6, the tangential (F_y) and normal (F_z) grinding force components values are depicted, in respect to different rake angle values. It should be noted that F_x values are not presented as there is no motion along this axis and force component values are negligible, compared to the two other force components. From Figure 6, it can be seen that both F_y and F_z values increase as rake angle values increase in magnitude and eventually, F_z values become greater than F_y values. The increasing trend for both force components is consistent with every relevant work in the literature [10,34–48]. Moreover, a slower increase is attested for F_y than for F_z, as F_z values increase 2.62 times, whereas F_y values increase only 1.14 times, as it was also observed in several works, such as [38,39,43,48].

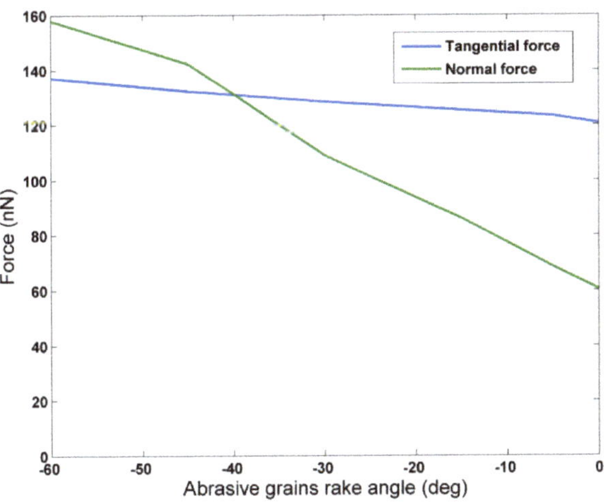

Figure 6. Grinding force components in respect to abrasive grains rake angle.

In Figure 7, the average friction coefficient values are depicted in respect to the rake angle of the abrasive grains. As in most relevant studies, the friction coefficient is calculated as the quotient of normal and tangential grinding force. From Figure 7, it can be seen that friction coefficient values increase as the magnitude of the rake angle increases. Friction coefficient value exceeds 1.0 above almost $-40°$ and eventually reaches the value 1.15 for rake angle value of $-60°$. In the relevant literature, several authors have also reported friction coefficient values over 1.0 in the case of negative rake angle values in nanocutting such as [38,39,43] and nanogrinding [10], with maximum values being 1.6, 1.4, 1.55, and 2.361, respectively. However, regarding the rake angle value above which the normal force exceeds the value of the tangential one, the researchers' findings do not always agree. Among the authors who reported values of normal force greater than the values of tangential force, Komanduri et al. [38] determined the critical rake angle value at almost −30° similar to Tong et al. [39], whereas Pei et al. [43] and Alhafez and Urbassek [46] calculated a value near −45° and Komanduri et al. [10] calculated a value of −15° in simulations of nanogrinding. Moreover, although Promyoo et al. [41,47] studied only positive values of rake angle over 0°, the maximum friction coefficient was found at 0° in both cases with a value of almost 0.4 and 0.886, respectively, with an increasing trend of friction coefficient as rake angle was decreased. Thus, the value predicted in the current work is in compliance with these works. On the other hand, Dai et al. [34,40] reported higher values of tangential force at least until −45° and Zhao et al. [36] reported a value of 0.55 at −40°. The reason for these variations is perhaps that friction coefficient is dependent on other process parameters to some extent, such as cutting speed, depth of cut, tool radius, and so, simulations conducted under different values of these parameters yield different results on the critical rake angle value which leads to $\mu > 1.0$.

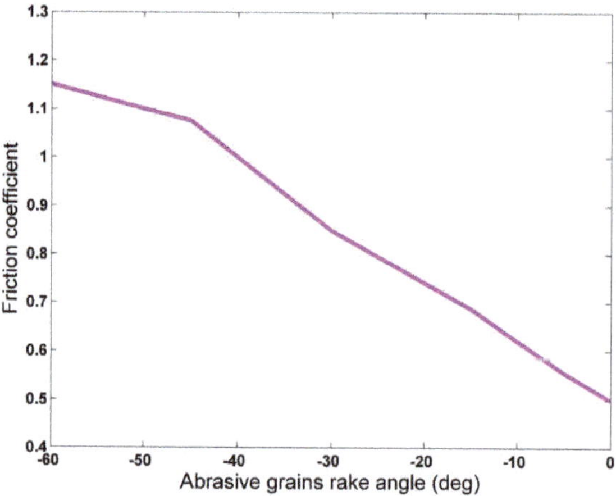

Figure 7. Average friction coefficient values in respect to abrasive grains rake angle.

Regarding the dimensions of the chip produced during nanogrinding, it is expected that different abrasive grain rake angle will produce significant alterations to the chip. In Figure 8, the chip morphology in the final snapshots of two representative cases of the second set of simulations, namely, at a rake angle value of 0° and −45° are compared, and in Figure 9, the height of the pile of the removed atoms is depicted in each case. It can be directly seen from Figure 9 that the height of the chip is considerably reduced for rake angle values up to −30° and then it remains practically stable. The reduction of chip height, with higher negative rake angle was also observed in several works in the relevant literature [34–38,40,43] due to the increased compression of the removed atoms by the abrasive grain face which contacts them at a highly inclined configuration.

In order to investigate the rake angle effect on subsurface damage, as well, analysis was performed by computing centrosymmetry parameter (CSP) values. According to CSP value for each atom of the Newtonian region of atoms, these atoms can be classified into five categories: for CSP < 3, the atoms belong to the original Face-centered Cubic (FCC) crystal; for 3 < CSP < 5, the atoms belong to a partial dislocation; for 5 < CSP < 8, the atoms belong to a stacking fault; and higher values of CSP indicate surface and surface edge atoms [49].

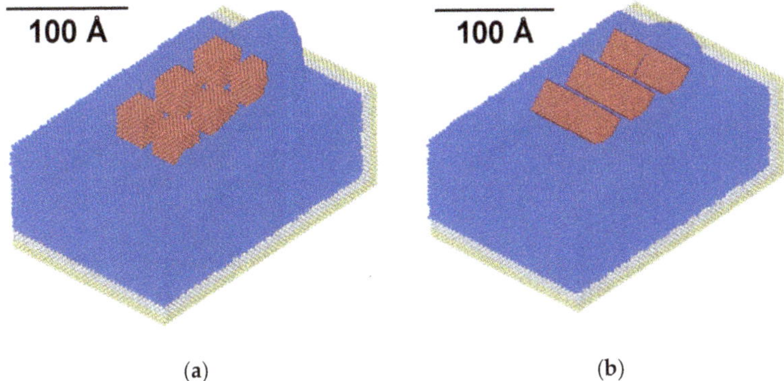

Figure 8. Snapshots of nanogrinding with abrasive grain rake angle of: (**a**) 0° and (**b**) −45°.

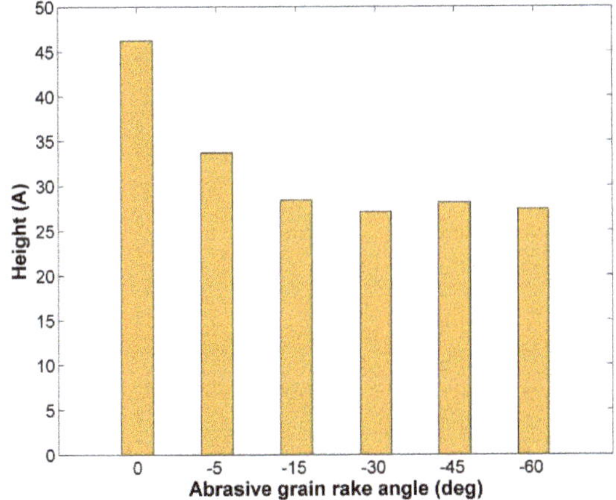

Figure 9. Chip height during nanogrinding in respect to abrasive grain rake angle.

The results depicted in Figure 10 indicate that an increase of abrasive grains rake angle leads gradually to more atoms of the workpiece being part of stacking faults or partial dislocations. Thus, grinding with highly negative rake angle grains leads to more intense plastic deformation of the workpiece as the imposed pressure on it becomes larger and material removal becomes gradually more difficult. A few works in the relevant literature also indicate this trend in nanocutting processes. Dai et al. [34,40] observed variations in the number of atoms pertinent to intrinsic stacking faults [34] during nanocutting of copper or in the number of atoms with different coordination during nanocutting of silicon [40]. Lai et al. [37] found out that plastic deformation was more intense in the workpiece with a decrease of rake angle towards highly negative values and that a lot of dislocations were concentrated in the subsurface area for these rake angle values. Similarly, Zhang et al. [45] computed higher dislocation densities for negative rake angles than for positive ones and Alhafez and Urbassek [46] identified a clear change of mechanism of plasticity at high negative rake angles, which was also related to the suppression of chip formation by thoroughly studying the dislocation movement.

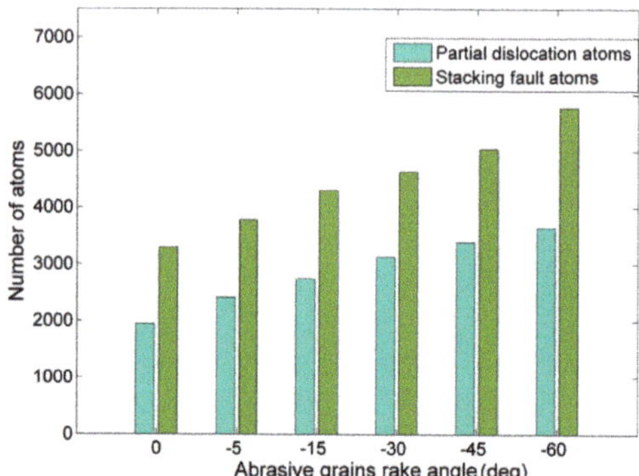

Figure 10. Number of atoms pertinent to partial dislocations and stacking faults in respect to abrasive grains rake angle.

4. Conclusions

In the present work, an MD model for nanogrinding with multiple abrasive grains was developed in order to investigate the effect of spacing between the adjacent rows of abrasive grains on ground surface and chip formation as well as the effect of abrasive grains rake angle on grinding forces, chip formation, and subsurface alterations. From the analysis of the results, several conclusions were drawn.

Regarding the effect of spacing of abrasive grains, it was shown that it affects considerably the formation of the ground surface and chip. Narrower spacing leads to the formation of a single groove in each pass and chip height is large, as chip is mainly concentrated in front of the abrasive grains. As spacing increases, due to more material remaining in the area between the rows of the abrasive grains, two separate grooves are formed with a common pileup and finally, there is even no interaction between the atoms removed from each groove, as separate pileups are formed. Chip height and average temperature measurements exhibit trends compatible with these observations and thus, support the qualitative findings.

Regarding the effect of abrasive grain rake angle, according to the findings, a direct correlation between grinding forces and rake angle was determined, with both tangential and normal grinding force increasing with higher negative rake angle values. The normal component of grinding force was more affected by the rake angle as it increased almost threefold when rake angle was −60° in comparison to 0°. At the same time, the friction coefficient increased to values over 1.0 for rake angles higher than −40° as higher friction forces occur to the more intense contact of abrasive grains and workpiece surface. Finally, analysis of centrosymmetry parameter values indicated an increase of plastic deformation in the workpiece for higher negative rake angle values, probably due to the higher pressure exerted on the workpiece surface. These results are important as they are consistent with trends observed also in the micro- and macroscale grinding and can provide an insight into phenomena which either cannot be observed in the experiments or cannot be captured with continuum method such as FEM.

Author Contributions: Conceptualization, N.E.K. and A.P.M.; data curation, N.E.K.; formal analysis, N.E.K.; funding acquisition, A.P.M.; investigation, N.E.K.; methodology, N.E.K. and A.P.M.; project administration, A.P.M.; resources, N.E.K. and A.P.M.; software, N.E.K.; supervision, A.P.M.; visualization, N.E.K.; writing—original draft, N.E.K.; and writing—review and editing, A.P.M. All authors have read and agreed to the published version of the manuscript.

Funding: This research received no external funding.

Conflicts of Interest: The authors declare no conflicts of interest.

References

1. Eder, S.J.; Cihak-Bayr, U.; Vernes, A.; Betz, G. Evolution of topography and material removal during nanoscale grinding. *J. Phys. D Appl. Phys.* **2015**, *48*, 465308. [CrossRef]
2. Goel, S. The current understanding on the diamond machining of silicon carbide. *J. Phys. D Appl. Phys.* **2014**, *47*, 243001. [CrossRef]
3. Cui, D.D.; Zhang, L.C. Nano-machining of materials: Understanding the process through molecular dynamics simulation. *Adv. Manuf.* **2017**, *5*, 20–34. [CrossRef]
4. Alder, B.; Wainwright, T. Studies in molecular dynamics. I. General method. *J. Chem. Phys.* **1959**, *31*, 459–466. [CrossRef]
5. Alder, B.; Wainwright, T. Studies in molecular dynamics. II. Behavior of a small number of elastic spheres. *J. Chem. Phys.* **1960**, *33*, 1439–1451. [CrossRef]
6. Belak, J.; Stowers, I.F. A molecular dynamics model of the orthogonal cutting process. In Proceedings of the ASPE Annual Conference, Rochester, NY, USA, 23–28 September 1990; p. 76.
7. Belak, J.; Boercker, D.B.; Stowers, I.F. Simulation of nanometer-scale deformation of metallic and ceramic surfaces. *MRS Bull.* **1993**, *21*, 55–60. [CrossRef]
8. Rentsch, R.; Inasaki, I. Molecular Dynamics Simulation for Abrasive processes. *CIRP Ann.* **1994**, *43*, 327–330. [CrossRef]
9. Rentsch, R.; Inasaki, I. Investigation of surface integrity by molecular dynamics simulation. *CIRP Ann.* **1995**, *44*, 295–298. [CrossRef]
10. Komanduri, R.; Chandrasekaran, N.; Raff, L.M. Some aspects of machining with negative-rake tools simulating grinding: A molecular dynamics simulation approach. *Philos. Mag. Part B* **1999**, *79*, 955–968. [CrossRef]
11. Lin, B.; Yu, S.Y.; Wang, S.X. An experimental study on molecular dynamics simulation in nanometer grinding. *J. Mater. Process Technol.* **2003**, *138*, 484–488. [CrossRef]
12. Li, J.; Fang, Q.; Liu, Y.; Zhang, L. A molecular dynamics investigation into the mechanisms of subsurface damage and material removal of monocrystalline copper subjected to nanoscale high speed grinding. *Appl. Surf. Sci.* **2014**, *303*, 331–343. [CrossRef]
13. Shimizu, J.; Zhou, L.B.; Eda, H. Simulation and experimental analysis of super high-speed grinding of ductile material. *J. Mater. Process Technol.* **2002**, *129*, 19–24. [CrossRef]
14. Li, J.; Fang, Q.; Zhang, L.; Liu, Y. Subsurface damage mechanism of high speed grinding process in single crystal silicon revealed by atomistic simulations. *Appl. Surf. Sci.* **2014**, *303*, 331–343. [CrossRef]
15. Guo, X.; Li, Q.; Liu, T.; Zhai, C.; Kang, R.; Jin, Z. Molecular dynamics study on the thickness of damage layer in multiple grinding of monocrystalline silicon. *Mater. Sci. Semicond. Process* **2016**, *51*, 15–19. [CrossRef]
16. Chen, P.; Zhang, Z.; An, T.; Yu, H.; Qin, F. Generation and distribution of residual stress during nano-grinding of monocrystalline silicon. *Jpn. J. Appl. Phys.* **2018**, *57*, 121302. [CrossRef]
17. Ren, J.; Hao, M.; Lv, M.; Wang, S.; Zhu, B. Molecular dynamics research on ultra-high-speed grinding mechanism of monocrystalline nickel. *Appl. Surf. Sci.* **2018**, *455*, 629–634. [CrossRef]
18. Liu, Y.; Zheng, Y.; Li, B. Investigation of high-speed nanogrinding mechanism based on molecular dynamics. In Proceedings of the ASME 2018 13th International Manufacturing Science and Engineering Conference MSEC2018, College Station, TX, USA, 18–22 June 2018. V004T03A024.
19. Liang, S.W.; Wang, C.H.; Fang, T.H. Rolling Resistance and Mechanical Properties of Grinded Copper Surfaces using Molecular Dynamics Simulation. *Nanoscale Res. Lett.* **2016**, *11*, 401. [CrossRef]
20. Zhang, Z.; Chen, P.; Qin, F.; An, T.; Yu, H. Mechanical properties of silicon in subsurface damage layer from nano-grinding studied by atomistic simulation. *AIP Adv.* **2018**, *8*, 055223. [CrossRef]
21. Xu, Y.; Wang, M.; Zhu, F.; Liu, X.; Liu, Y.; He, L. Study on subsurface damage of wafer silicon containing through silicon via in thinning. *Eur. Phys. J. Plus* **2019**, *134*, 234. [CrossRef]
22. Ren, J.; Hao, M.; Liang, G.; Wang, S.; Lv, M. Study of subsurface damage of monocrystalline nickel in nanometric grinding with spherical abrasive grain. *Phys. B Condens. Matter* **2019**, *560*, 60–66. [CrossRef]
23. Ren, J.; Liang, G.; Lv, M. Effect of different crystal orientations on the surface integrity during nanogrinding of monocrystalline nickel. *Model. Simul. Mater. Sci. Eng.* **2019**, *27*, 075007. [CrossRef]
24. Liu, Y.; Li, B.; Kong, L. Atomistic insights on the nanoscale single grain scratching mechanism of silicon carbide ceramic based on molecular dynamics simulation. *AIP Adv.* **2018**, *8*, 035109. [CrossRef]

25. Xu, Y.; Zhu, F.; Wang, M.; Liu, X.; Liu, S. Molecular Dynamics Simulation on Grinding Process of Cu-Si and Cu-SiO$_2$ Composite Structures. In Proceedings of the 2018 19th International Conference on Electronic Packaging Technology, Shanghai, China, 8–11 August 2018; pp. 79–83.

26. Wang, Q.; Fang, Q.; Li, J.; Tian, Y.; Liu, Y. Subsurface damage and material removal of Al-Si bilayers under high-speed grinding using molecular dynamics (MD) simulation. *Appl. Phys. A* **2019**, *125*, 514. [CrossRef]

27. Fang, Q.; Wang, Q.; Li, J.; Zeng, X.; Liu, W. Mechanisms of subsurface damage and material removal during high speed grinding processes in Ni/Cu multilayers using a molecular dynamics study. *RSC Adv.* **2017**, *7*, 42047. [CrossRef]

28. Xu, Y.; Wang, M.; Zhu, F.; Liu, X.; Chen, Q.; Hu, J.; Lu, Z.; Zeng, P.; Liu, Y. A molecular dynamic study of nano-grinding of a monocrystalline copper-silicon substrate. *Appl. Surf. Sci.* **2019**, *493*, 933–947. [CrossRef]

29. Fang, X.; Kang, Q.; Ding, J.; Sun, L.; Maeda, R.; Jiang, Z. Stress distribution in silicon subjected to atomic scale grinding with a curved tool path. *Materials* **2020**, *13*, 1710. [CrossRef]

30. Eder, S.J.; Bianchi, D.; Cihak-Bayr, U.; Vernes, A.; Betz, G. An analysis method for atomistic abrasion simulations featuring rough surfaces and multiple abrasive particles. *Comput. Phys. Commun.* **2014**, *185*, 2456–2466. [CrossRef]

31. Eder, S.J.; Cihak-Bayr, U.; Pauschitz, A. Nanotribological simulations of multi-grit polishing and grinding. *Wear* **2015**, *340*, 25–30. [CrossRef]

32. Adams, J.B.; Foiles, S.M.; Wolfer, W.G. Self-diffusion and impurity diffusion of fcc metals using the five-frequency model and the Embedded Atom Method. *J. Mater. Res.* **1989**, *4*, 102–112. [CrossRef]

33. Fang, T.H.; Weng, C.L. Three-dimensional molecular dynamics analysis of processing using a pin tool on the atomic scale. *Nanotechnology* **2000**, *11*, 148–153. [CrossRef]

34. Dai, H.; Du, H.; Chen, J.; Chen, G. Investigation of tool geometry in nanoscale cutting single-crystal copper by molecular dynamics simulation. *Proc. Inst. Mech. Eng. J.* **2019**, *233*, 1208–1220. [CrossRef]

35. Ge, J.H.; Zhang, C.H.; Wang, Y.P.; Sui, X.L.; Guo, Y.B. Simulation analysis of the effects of tool rake angle for workpiece temperature in single crystal copper nanometric cutting process. *Int. J. Hybrid Inf. Technol.* **2016**, *9*, 407–414.

36. Zhao, H.; Zhang, L.; Zhang, P.; Shi, C. Influence of geometry in nanometric cutting single-crystal copper via MD simulation. *Adv. Mater. Res.* **2012**, *421*, 123–128. [CrossRef]

37. Lai, M.; Zhang, X.D.; Fang, F.Z. Study on critical rake angle in nanometric cutting. *Appl. Phys. A* **2012**, *108*, 809–818. [CrossRef]

38. Komanduri, R.; Chandrasekaran, N.; Raff, L.M. Molecular dynamics simulation of the nanometric cutting of silicon. *Philos. Mag. Part B* **2001**, *81*, 1989–2019. [CrossRef]

39. Tang, Y.; Liu, Q.; Wu, Y.; Zhang, K. Generation mechanism of micro/nano machined surfaces based on molecular dynamics. *Adv. Mat. Res.* **2010**, *97*, 3104–3107. [CrossRef]

40. Dai, H.; Chen, G.; Fang, Q.; Yin, J. The effect of tool geometry on subsurface damage and material removal in nanometric cutting single-crystal silicon by a molecular dynamics simulation. *Appl. Phys. A* **2016**, *122*, 804. [CrossRef]

41. Promyoo, R.; El-Mounayri, H.; Yang, X. Molecular dynamics simulation of nanometric cutting. *Mach. Sci. Technol.* **2010**, *14*, 423–439. [CrossRef]

42. Shi, J.; Wang, Y.; Yang, X. Nano-scale machining of polycrystalline coppers—Effects of grain size and machining parameters. *Nanoscale Res. Lett.* **2013**, *8*, 500. [CrossRef]

43. Pei, Q.X.; Lu, C.; Fang, F.Z.; Wu, H. Nanometric cutting of copper: A molecular dynamics study. *Comput. Mater. Sci.* **2006**, *37*, 434–441. [CrossRef]

44. Ye, Y.Y.; Biswas, R.; Morris, J.R.; Bastawros, A.; Chandra, A. Molecular dynamics simulation of nanoscale machining of copper. *Nanotechnology* **2003**, *14*, 390–396. [CrossRef]

45. Zhang, J.; Zheng, H.; Shuai, M.; Li, Y.; Yang, Y.; Sun, T. Molecular dynamics modeling and simulation of diamond cutting of cerium. *Nanoscale Res. Lett.* **2017**, *12*, 464. [CrossRef]

46. Alhafez, I.A.; Urbassek, H.M. Influence of the rake angle on nanocutting of Fe single crystals: A molecular dynamics study. *Crystals* **2020**, *10*, 516. [CrossRef]

47. Promyoo, R.; El-Mounayri, H.; Yang, X. Molecular dynamics simulation of nanometric machining under realistic cutting conditions. In Proceedings of the ASME 2008 International Manufacturing Science and Engineering Conference, Evanston, IL, USA, 7–10 October 2008; pp. 235–243.

48. Ji, C.; Shi, J.; Wang, Y.; Liu, Z. A numeric investigation of friction behaviors along tool/chip interface in nanometric machining of a single crystal copper structure. *Int. J. Adv. Manuf. Technol.* **2013**, *68*, 365–374. [CrossRef]
49. Tong, Z.; Liang, Y.; Jiang, X.; Luo, X. An atomistic investigation on the mechanism of machining nanostructures when using single tip and multi-tip diamond tools. *Appl. Surf. Sci.* **2014**, *290*, 458–465. [CrossRef]

 *micromachines*

Article

A Method to Determine the Minimum Chip Thickness during Longitudinal Turning

Michal Skrzyniarz

Department of Manufacturing Engineering and Metrology, Kielce University of Technology, 25-314 Kielce, Poland; mskrzyniarz@tu.kielce.pl; Tel.: +48-4134-24-417

Received: 29 October 2020; Accepted: 19 November 2020; Published: 24 November 2020

Abstract: Micromachining, which is used for various industrial purposes, requires the depth of cut and feed to be expressed in micrometers. Appropriate stock allowance and cutting conditions need to be selected to ensure that excess material is removed in the form of chips. To calculate the allowance, it is essential to take into account the tool nose radius, as this cutting parameter affects the minimum chip thickness. Theoretical and numerical studies on the topic predominate over experimental ones. This article describes a method and a test setup for determining the minimum chip thickness during turning. The workpiece was ground before turning to prevent radial runout and easily identify the transition zone. Contact and non-contact profilometers were used to measure surface profiles. The main aim of this study was to determine the tool–workpiece interaction stages and the cutting conditions under which material was removed as chips. Additionally, it was necessary to analyze how the feed, cutting speed, and edge radius influenced the minimum chip thickness. This parameter was found to be dependent on the depth of cut and feed. Elastic and plastic deformation and ploughing were observed when the feed rate was lower than the cutting edge radius.

Keywords: turning; minimum chip thickness; micromachining

1. Introduction

The manufacture of precision machine parts for the electrical, electronic, computer, biotechnological, and machine tool industries requires an understanding of the physical phenomena occurring at the microscale [1,2], especially when very low-depth cuts are involved [3,4]. One of the key factors affecting the surface quality in micromachining is the minimum chip thickness [5,6]. This parameter defines the minimum depth of cut at which, under specific cutting conditions, material is removed in the form of chip [7,8].

There are three stages of interaction between the tool and the workpiece [9]: (I) elastic and plastic deformation of the workpiece at $a_p < h_{min}$; (II) elastic and plastic deformation and ploughing at $a_p = h_{min}$; (III) shearing (chip formation) at $a_p > h_{min}$.

Researchers dealing with the minimum chip thickness approach the problem both theoretically and experimentally [10,11]. The depth of cut and other cutting parameters are used to optimize the process.

Currently, there are many mathematical models used to predict h_{min}. The most important formulae are summarized in Table 1 [12–18].

Table 1. Models used to calculate the minimum chip thickness.

No.	Model	Reference
1.	$h_{min} = r_n(1 - \cos\theta)$	[12]
2.	$h_{min} = k\, r_n$	[13]
3.	$h_{min} \geq 0.5\, r_n\left[1 - \frac{2\tau_a}{Y_{ch}}\right]$	[14]
4.	$h_{min} = r_n\left(1 - \frac{Fy+\mu Fx}{\sqrt{(Fx^2+Fy^2)(1+\mu^2)}}\right)$	[15]
5.	$h_{min} = r_n\left(1 - \cos\left(\frac{\pi}{4} - \frac{\beta}{2}\right)\right)$	[16,17]
6.	$h_{min} = r_n\left[1 - \cos\left(arc\, ctg\frac{A_f}{A_c}\right)\right]$	[18]

The calculation of the minimum chip thickness according to Formula (1) requires the stagnation point (A) to be determined, which is defined by the stagnant angle (θ) in relation to the edge radius (r_n). The stagnation point is used to determine the zones of interaction between the tool and the workpiece. When the thickness of the material to be removed is below the stagnation point, the cutting process involves ploughing with elastic and plastic deformation of the workpiece surface; however, when the thickness of the material to be removed is above the stagnation point, chip removal occurs [12].

In [13], Kawalec proposes a formula (Formula (2)) in which the minimum chip thickness (h_{min}) depends on the edge radius (r_n) and the coefficient of friction between the tool and the workpiece (k), with the latter ranging from 0.1 to 1. Other studies confirm that the coefficient of friction affects the surface roughness [19].

In [14], Grzesik applies the molecular–mechanical theory of friction to formulate a model to predict the minimum chip thickness (3). The strain hardening curve is used to calculate the equivalent strain and shear strength of the material in the adhesive junction. In [20], Grzesik analyzes the phenomena responsible for both partial material separation and chip formation.

Formula (4) proposed by Yuan et al. assumes that h_{min} is dependent on the edge radius (r_n), the coefficient of friction between the tool and the workpiece (μ), and the force ratio F_y/F_x [15].

In [16,17], the model for h_{min} (5) assumes that the stagnation point (A), based on the stagnant angle (θ), is affected by the edge radius (r_n). When the stagnant angle is 45°, the minimum chip thickness ranges between 0.2 and 0.35 r_n.

The major assumption of model (6) is to maintain constant temperature conditions irrespective of the cutting conditions and to measure the cutting forces that affect the directional coefficients A_f and A_c [18].

The approach proposed by L'vov suggests that $h_{min} = 0.29\, r_n$ at $\theta = 37.6°$ [21]. In [22], however, h_{min} is reported to be 0.21 r_n. Yuan et al. show that for aluminum alloys, the parameter h_{min} ranges from 0.25 to 0.32 r_n [15]. A similar relationship is described in [23], where the minimum chip thickness is 0.25–0.3 r_n.

Although most studies on the minimum chip thickness deal with mathematical models, some researchers approach the problem practically [24]. In [25], the method involves preparing a conical sample with a taper angle of 5°, cutting it along the cone height, and placing the two halves joined end-to-end between the centers. The sample is first ground and then turned. After the cutting operations, the halves are separated and the tool–workpiece interaction zones are analyzed using a workshop microscope.

Other methods used to determine the minimum chip thickness require the application of acoustic emission signals to provide information on the phenomena taking place in the cutting zone, especially material separation [26,27].

The literature on the subject does not mention any practical methods for determining the minimum chip thickness during turning, nor does it analyze the tool–workpiece interaction in relation to the minimum chip thickness based on surface profiles. The aim of this study was to develop an experimental method to determine h_{min} during longitudinal turning.

2. New Method and Test Setup for Determining the Minimum Chip Thickness during Turning

The experimental method for determining h_{min} presented here was developed to overcome the limitations of the approach presented in [25], mainly because of the time- and labor-consuming sample preparation process and the discontinuous cutting, which was interrupted at the point of contact of the two sample halves. The new method uses a conical sample with known convergence ($C = 0.008$). The workpiece is mounted on a mandrel and held in place with a lathe dog at one end and a nut at the other, so there is no need to change the workpiece holding devices. The workpiece system is supported between centers. Grinding is used to produce a tapered surface; it also helps in analyzing the tool–workpiece interactions and to prevent radial run-out during turning so that the material is removed at a constant cut depth around the circumference. The torque is transmitted to the workpiece through the lathe dog. The workpiece holding system used to determine the minimum chip thickness is shown in Figure 1. The turning was performed on a DMG CTX Alpha 500 CNC turning and milling machine.

Figure 1. Workpiece holding system used to determine the minimum chip thickness during turning.

The method proposed here does not require the cutting process to be interrupted; turning is performed as a continuous process. The sample preparation process is quick and simple. The use of grinding allows us to easily indicate the initial stage of interaction between the tool and the workpiece during turning. The method, however, requires that a concentric workpiece be produced; this means turning needs to be performed at the same cut depth around the circumference.

After turning, the workpiece is analyzed using a contact skidless profilometer (TOPO 01). The primary profile is used as the basis for determining the stages of the tool–workpiece interaction. The sections with regular surface profiles indicate that turning with chip formation has taken place. Profile irregularity suggests that turning with no chip formation has taken place (ploughing). The results obtained in the experiment were accurate and repeatable because the profilometer used is a high-resolution profilometer. The experimental findings were confirmed by 3D surface measurements with a Talysurf CCI optical profilometer.

3. Experiment and Results

The experiments were carried out using samples made of X37CrMoV5-1 steel. The material hardness before grinding was 34.9 ± 5 HRC. A Sandvik Coromant CoroTurn 107 SDJCL 2020K 11 turning tool holder was used for longitudinal turning. The inserts tested were 55° diamond-shaped positive turning inserts with a 7° clearance angle and a 93° entering angle. Although manufacturers of cutting tools do not provide information on the edge radius in their product catalogs, this parameter needs to be taken into consideration in micromachining. Three types of standard turning inserts were used in the tests: DCMT 11 T3 08–PF 4325, DCMT 11 T3 08–MF 1105, and DCGT 11 T3 08–UM 1125. The edge radius of each insert was measured before cutting.

The edge radii were measured using a special setup employed by Sandvik. The measurement involved moving the measuring tip around the cutting edge and recording the data for each position. The edge radius was calculated using the device software. An example of an edge radius measurement report is shown in Figure 2.

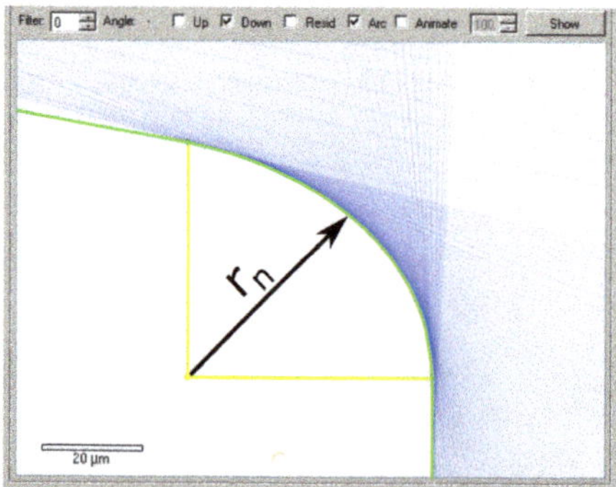

Figure 2. Measurement of the edge radius (r_n) of the DCMT 11 T3 08–PF 4325 insert.

Ten inserts were measured for each type of insert to determine their edge radii. For the DCMT 11 T3 08–PF 4325 type, the edge radius values ranged between 44.4 and 55.9 µm; for the DCMT 11 T3 08–MF 1105 type, its values ranged from 29.6 to 38.9 µm; for the DCGT 11 T3 08–UM 1125 type, the values ranged from 5.3 to 11.5 µm.

After turning, the workpiece surface was analyzed with a contact skidless profilometer (TOPO 01). Primary profiles obtained for specific cutting conditions were the basis for determining the minimum chip thickness. Figure 3 shows how the parameter was determined for longitudinal turning.

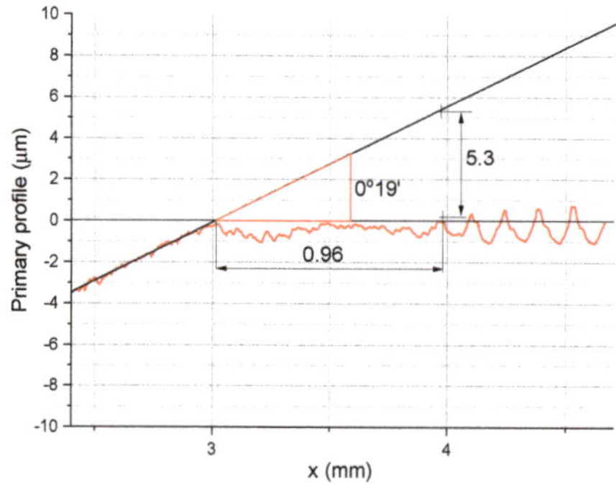

Figure 3. Determining h_{min} from a surface profile obtained with a contact profilometer (v_c = 360 m/min, f = 0.03 mm/rev, and r_n = 49.9 µm).

As can be seen from Figure 3, there are three characteristic zones: the grinding zone, transition zone, and turning zone. In zone 1, there is no interaction between the turning tool and the workpiece. In zone 2, the material undergoes elastic and plastic deformation; partial removal of material also occurs. The irregularity of the profile in this zone indicates an unstable cutting process. The zone is approximately 0.96 mm in length. In order to determine the minimum chip thickness for which chip removal occurred, the ground surface profile was extended as a straight line to find the point indicating the end of zone 2. In the case considered here, i.e., for v_c = 360 m/min, f = 0.03 mm/rev, and r_n = 49.9 µm, the minimum chip thickness was 5.3 µm. The regular profile in the third zone represents turning with chip formation.

Figure 4 shows the surface profiles obtained at v_c = 360 m/min, f = 0.03 mm/rev, and r_n = 49.9 µm.

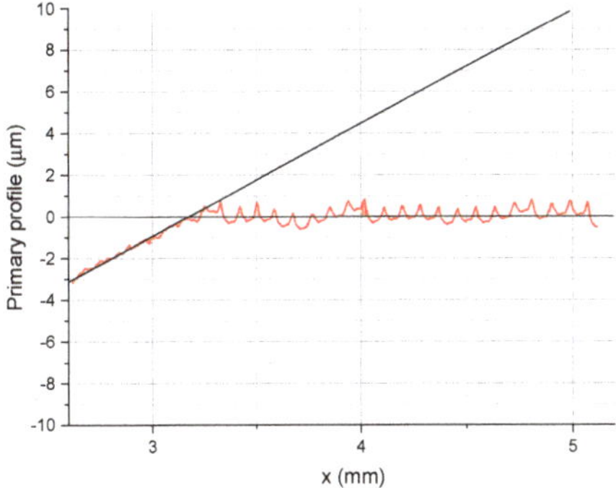

Figure 4. Surface profiles obtained with a contact profilometer for a workpiece machined at v_c = 360 m/min, f = 0.09 mm/rev, and r_n = 49.9 µm.

From the profile in Figure 4, it is clear that the material did not undergo elastic and plastic deformation, nor was it separated partially (ploughing). Partial removal of material was observed at f = 0.03 mm/rev. At feed rates higher than 0.06 mm/rev there was no transition zone between grinding and turning. No unstable cutting was reported.

When the cutting was performed using a DMCT 11 T3 08–PF 4325 insert with an edge radius of 49.9 µm, elastic and plastic deformation occurred. With feed rates lower than the edge radius r_n, partial removal of material (ploughing) was observed. Feed rates greater than the edge radius ensured material removal in the form of chips. To verify the validity of the method, the experiment was carried out with feed rates ranging from 0.01 mm/rev to 0.06 mm/rev. Table 2 shows the cutting parameters used to determine the minimum chip thickness. The data provided in Table 2 indicate that the unstable cutting process responsible for elastic and plastic deformation and partial material removal (ploughing) was observed with feed rates smaller than the edge radius.

Table 2. Cutting conditions for the three insert types.

No.	v_c [m/min]	f [mm/rev]	n [rev/min]	h_{min} [µm]	No.	v_c [m/min]	f (mm/rev)	n [rev/min]	h_{min} [µm]
	DCMT 11 T3 08–PF 4325, r_n = 49.9 µm								
1.	360	0.01	2349	8.37 ± 0.87	7.	330	0.03	2167	6.24 ± 0.71
2.	360	0.02	2349	6.07 ± 1.39	8.	345	0.03	2265	5.95 ± 1.21
3.	360	0.03	2349	5.47 ± 1.39	9.	360	0.03	2364	5.74 ±1.19
4.	360	0.04	2349	3.61 ± 1.21	10.	375	0.03	2462	4.98 ± 2.27
5.	360	0.05	2349	———	11.	390	0.03	2561	5.77 ± 0.89
6.	360	0.06	2349	———	12.	405	0.03	2659	9.24 ± 2.63
					13.	420	0.03	2758	10.79 ± 1.35
	DCMT 11 T3 08–MF 1105, r_n = 30 µm					DCGT 11 T3 08–UM 1125, r_n = 7.8 µm			
14.	360	0.01	2349	20.56 ± 5.37	20.	360	0.01	2349	———
15.	360	0.02	2349	6.93 ± 3.81	21.	360	0.02	2349	———
16.	360	0.03	2349	———	22.	360	0.03	2349	———
17.	360	0.04	2349	———	23.	360	0.04	2349	———
18.	360	0.05	2349	———	24.	360	0.05	2349	———
19.	360	0.06	2349	———	25.	360	0.06	2349	———

The data given in Table 2 were then used to illustrate how the parameter h_{min} was dependent on the feed rate and cutting speed. The relationship between h_{min} and the feed rate for the DCMT 11 T3 08–PF 4325 insert with a 49.9 μm edge radius is shown in Figure 5, while the relationship between h_{min} and the cutting speed for the same insert type is given in Figure 6. When longitudinal turning was performed using a DCMT 11 T3 08–MF 1105 insert with an edge radius of 30 μm, the parameter h_{min} was read only at feed rates lower than 0.02 mm/rev. Since no characteristic zones were identified at the feeds tested for the DCGT 11 T3 08–UM 1125 inserts with $r_n = 7.8$ μm, this type of insert will not be discussed further in this article.

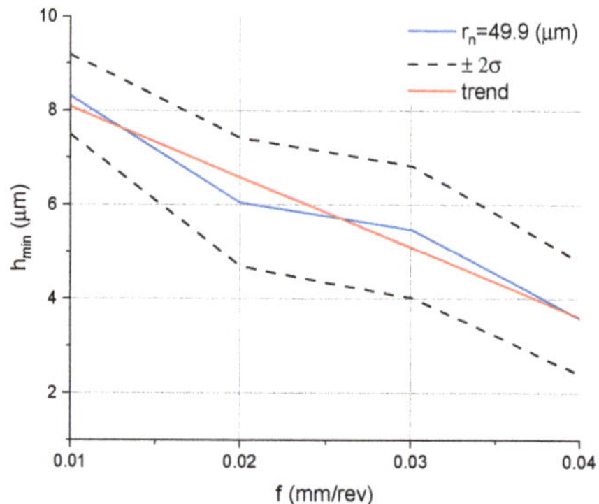

Figure 5. Minimum chip thickness vs. feed for the DCMT 11 T3 08–PF 4325 insert with a 49.9 μm edge radius.

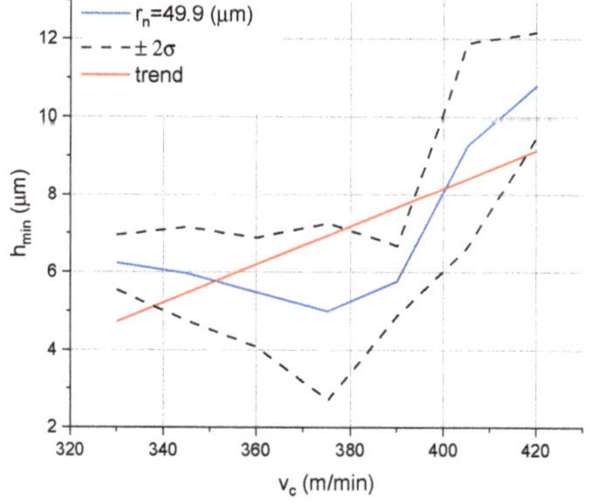

Figure 6. Minimum chip thickness vs. cutting speed for the DCMT 11 T3 08–PF 4325 insert with a 49.9 μm edge radius.

From the results provided in Figure 5, it is evident that h_{min} decreases with decreasing feed rate. The parameter could not be read at feed rates greater than the edge radius. Low values for the feed rate, depth of cut, and edge radius resulted in ploughing and elastic and plastic deformation of the material that was removed. When the cutting parameters increased, the elastic and plastic deformation decreased and the material was removed as chips, with the chip thickness being equal to the depth of cut. A decrease in h_{min} with increasing feed rate was due to an increase in the cross-sectional area of the cut material.

Figure 6 shows the effect of the cutting speed on the parameter h_{min}. A decrease in h_{min} was reported at speeds ranging from 330 to 375 m/min; with a faster cutting speeds, the parameter increased. This phenomenon may have been due to an increase in the edge radius as a result of insert wear. Faster cutting speeds caused faster tool wear. It is important to note that one insert was used to test each insert type across the whole range of cutting speeds. In Figure 6, the red straight line representing the average values of the parameter h_{min} over the whole cutting speed range increases, which allows us to conclude that the minimum chip thickness increases with increasing cutting speed.

The test results confirm that the minimum chip thickness is dependent on the edge radius. The edge radius changes due to the wear of the tool associated with machining. The method for determining the minimum chip thickness proposed in this article requires analyzing the primary surface profile. The analysis takes into account the tool wear, as it is assumed to contribute to an increase in h_{min}. To ensure appropriate interpretation of the test results concerning the parameter h_{min}, it was essential to analyze the machined surfaces using an optical profilometer to identify the zones of the interaction between the tool and the workpiece. The data obtained for an insert measuring 49.9 μm are illustrated in Figures 7 and 8.

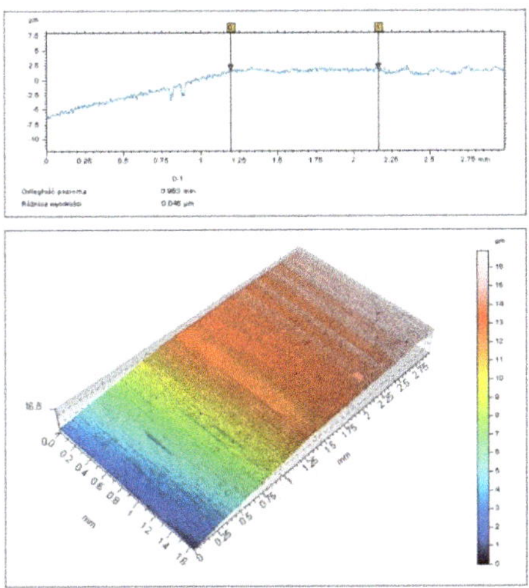

Figure 7. A 3D surface image and the corresponding 2D profile for a sample cut at v_c = 360 m/min and f = 0.03 mm/rev using an insert with an edge radius of 49.9 μm.

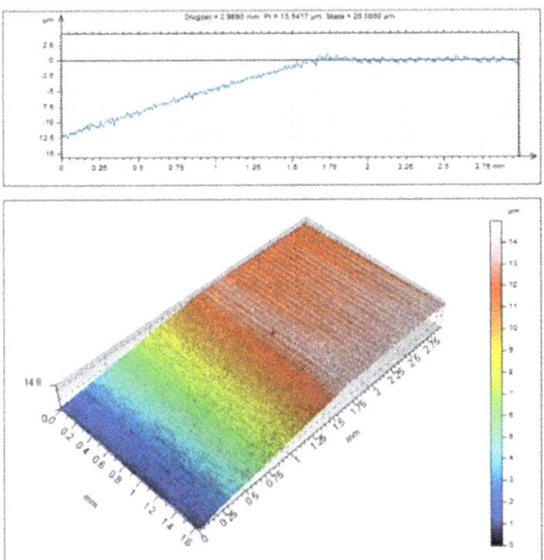

Figure 8. A 3D surface image and the corresponding 2D profile for a sample cut at v_c = 360 m/min and f = 0.09 mm/rev using an insert with an edge radius of 49.9 μm.

Figure 7 shows a 3D surface image and the corresponding 2D profile for a sample cut at a feed rate of 0.03 mm/rev using an insert with an edge radius of 49.9 μm. As can be seen, zone 1 represents the ground surface. The first interaction of the tool with the workpiece is clearly visible. The moment is marked as point 0 on the 2D profile. Then, up to point 1, no regular profile typical of turning with chip removal was observed. This suggests that the material undergoes elastic and plastic deformation and that ploughing occurs. The phenomena are also visible in the 3D image of the surface. The transition region is 0.96 mm in length, with the value being the same as that determined with a contact profilometer. The same analysis was carried out for the sample illustrated in Figure 4. The results of non-contact 3D measurements are shown in Figure 8.

Figures 4 and 8 reveal that turning at a feed rate of 0.09 mm/rev using a tool with an edge radius of 49.9 μm does not cause any visible elastic and plastic deformation. This is due to the fact that the first tool–workpiece interaction was sufficient for chip formation. The 2D profile shows that once the tool interacts with the workpiece, the profile becomes regular, which is characteristic of turning with chip removal.

As the chip thickness is largely dependent on the feed rate during turning, the minimum chip thickness must be defined as the minimum feed rate at which a chip is formed.

In most studies, h_{min} is assumed to be less than or equal to 0.4 of the edge radius. In this article, the minimum chip thickness is determined as the point on the profile where the irregularity ends. When the cutting speed and feed increase, the elastic and plastic deformation of the layer removed from the workpiece decreases. For the DCMT 11T3-MF 1105 insert, h_{min} was read at a maximum feed (0.02 mm/rev) and was 0.69 r_n; for the DCMT 11 T3 08 insert, the value of h_{min} read at a maximum feed (0.04 mm/rev) was 0.17 r_n. High values of the parameter h_{min} may have been due to the wear of the cutting edge (an increase in the edge radius), as wear is responsible for larger cutting edge and tool nose radii.

Comparing the values of the parameter h_{min} obtained in the experiment with those calculated using the different models available in the literature is difficult because it requires the determination of many constants used in these formulae. Additionally, some of the models are dedicated to specific materials or tool–material systems.

4. Conclusions

The following conclusions were drawn from the experimental results:

1. The method employed to measure h_{min} requires that a workpiece with a simple shape and a high-resolution instrument be used to measure surface profiles so that the characteristic tool–workpiece interaction zones can be defined easily and accurately;
2. The method overcomes the drawbacks of radial run-out during turning by supporting the workpiece on a mandrel;
3. A non-contact 3D optical profilometer can be sued as an alternative to the contact skidless profilometer used in the experiment;
4. The parameter h_{min} is dependent on the cutting edge geometry, depth of cut, and feed rate. For the DCMT 11T3–MF 1105 insert, h_{min} was read at a maximum feed (0.02 mm/rev) and was 0.69 r_n; for the DCMT 11 T3 08 insert, the value of h_{min} read at a maximum feed (0.04 mm/rev) was 0.17 r_n;
5. The parameter h_{min} was affected by the feed rate and cutting speed. For $0 < f < r_n$, an increase in the feed rate caused a decrease in the minimum chip thickness h_{min}'; higher cutting speeds were responsible for higher values of h_{min};
6. The parameter h_{min} can be used as an auxiliary parameter to determine the wear of the cutting edge. The minimum chip thickness increases with increasing edge radius;
7. The values of the parameter h_{min} were determined for samples made of X37CrMoV5-1 steel. They are likely to be different for other tool and workpiece materials, as hardness, whether of the workpiece or the tool, is a crucial parameter in cutting.

Funding: This research received no external funding.

Acknowledgments: The author gratefully acknowledges the assistance and materials provided for the experiment by Sandvik Polska Sp. z o.o.

Conflicts of Interest: The author declares no conflict of interest.

Nomenclature

v_c [m/min] cutting speed
f [mm/rev] feed rate
a_p [mm] depth of cut
n [rev/min] spindle speed
h_{min} [μm] minimum chip thickness
r_n [μm] cutting edge radius
r_ε [mm] tool nose radius
τ_a [MPa] shear strength in the adhesive junction
Y_{ch} [MPa] yield stress due to strain hardening
k [-] coefficient of friction
μ [-] coefficient of friction
F_x [N] horizontal force
F_y [N] vertical force
A_f [-] directional coefficients
A_c [-] directional coefficients
β [°] friction angle between the tool and the workpiece
π [°] friction angle
θ [°] stagnant angle

References

1. Nowakowski, L.; Skrzyniarz, M.; Miko, E. The assessment of the impact of the installation of cutting plates in the body of the cutter on the size of generated vibrations and the geometrical structure of the surface. *Eng. Mech.* **2017**, *2017*, 734–737.
2. Nowakowski, L.; Miko, E.; Skrzyniarz, M. Milling with a tool with unevenly distributed cutting plates. *Eng. Mech.* **2017**, *2017*, 726–729.
3. Yao, Y.; Zhu, H.; Huang, C.; Wang, J.; Zhang, P.; Yao, P.; Wang, X. Determination of the minimum chip thickness and the effect of the plowing depth on the residual stress field in micro-cutting of 18 Ni maraging steel. *Int. J. Adv. Manuf. Technol.* **2020**, *106*, 345–355. [CrossRef]
4. Sahoo, P.; Patra, K.; Szalay, T.; Dyakonov, A.A. Determination of minimum uncut chip thickness and size effects in micro-milling of P-20 die steel using surface quality and process signal parameters. *Int. J. Adv. Manuf. Technol.* **2020**, *106*, 4675–4691. [CrossRef]
5. Zmarzly, P.; Kozior, T.; Gogolewski, D. Dimensional and Shape Accuracy of Foundry Patterns Fabricated Through Photo-Curing. *Teh. Vjesn.* **2019**, *26*, 1576–1584. [CrossRef]
6. Przestacki, D.; Chwalczuk, T.; Wojciechowski, S. The study on minimum uncut chip thickness and cutting forces during laser-assisted turning of WC/NiCr clad layers. *Int. J. Adv. Manuf. Technol.* **2017**, *91*, 3887–3898. [CrossRef]
7. Nowakowski, L.; Skrzyniarz, M.; Miko, E. The analysis of relative oscillation during face milling. *Eng. Mech.* **2017**, *2017*, 730–733.
8. Wojciechowski, S.; Matuszak, M.; Powałka, B.; Madajewski, M.; Maruda, R.W.; Królczyk, G.M. Prediction of cutting forces during micro end milling considering chip thickness accumulation. *Int. J. Mach. Tools Manuf.* **2019**, *147*, 103466. [CrossRef]
9. Grzesik, W. *Podstawy Skrawania Materiałow Konstrukcyjnych*; Wyd. 2 zm. i rozsz; Wydawnictwa Naukowo-Techniczne: Warszawa, Poland, 2010; ISBN 8320436680.
10. Ortiz-de-Zarate, G.; Sela, A.; Saez-de-Buruaga, M.; Cuesta, M.; Madariaga, A.; Garay, A.; Arrazola, P.J. Methodology to establish a hybrid model for prediction of cutting forces and chip thickness in orthogonal cutting condition close to broaching. *Int. J. Adv. Manuf. Technol.* **2019**, *101*, 1357–1374. [CrossRef]

11. Abdulkadir, L.N.; Abou-El-Hossein, K. Observed edge radius behavior during MD nanomachining of silicon at a high uncut chip thickness. *Int. J. Adv. Manuf. Technol.* **2019**, *101*, 1741–1757. [CrossRef]

12. Malekian, M.; Mostofa, M.G.; Park, S.S.; Jun, M.B.G. Modeling of minimum uncut chip thickness in micro machining of aluminum. *J. Mater. Process. Technol.* **2012**, *212*, 553–559. [CrossRef]

13. Kawalec, M. *Fizyczne i Technologiczne Zagadnienia Przy Obróbce z Małymi Grubościami Warstwy Skrawanej*; Wydawnictwa Politechniki Poznańskiej: Poznań, Poland, 1980.

14. Grzesik, W. A revised model for predicting surface roughness in turning. *Wear* **1996**, *194*, 143–148. [CrossRef]

15. Yuan, Z.J.; Zhou, M.; Dong, S. Effect of diamond tool sharpness on minimum cutting thickness and cutting surface integrity in ultraprecision machining. *J. Mater. Process. Technol.* **1996**, *62*, 327–330. [CrossRef]

16. Son, S.; Lim, H.; Ahn, J. The effect of vibration cutting on minimum cutting thickness. *Int. J. Mach. Tools Manuf.* **2006**, *46*, 2066–2072. [CrossRef]

17. Son, S.M.; Lim, H.S.; Ahn, J.H. Effects of the friction coefficient on the minimum cutting thickness in micro cutting. *J. Mater. Process. Technol.* **2005**, *45*, 529–535. [CrossRef]

18. Storch, B.; Zawada-Tomkiewicz, A. Distribution of unit forces on the tool edge rounding in the case of finishing turning. *Int. J. Adv. Manuf. Technol.* **2012**, *60*, 453–461. [CrossRef]

19. Leksycki, K.; Feldshtein, E.; Królczyk, G.M.; Legutko, S. On the Chip Shaping and Surface Topography When Finish Cutting 17-4 PH Precipitation-Hardening Stainless Steel under Near-Dry Cutting Conditions. *Materials* **2020**, *13*, 2188. [CrossRef]

20. Grzesik, W. A real picture of plastic deformation concentrated in the chip produced by continuous straight-edged oblique cutting. *J. Mater. Process. Technol.* **1991**, *31*, 329–344. [CrossRef]

21. L'vov, N.P. Determining the minimum possible chip thickness. *Mach. Tool.* **1969**, *4*, 40–45.

22. Basuray, P.K.; Misra, B.K.; Lal, G.K. Transition from ploughing to cutting during machining with blunt tools. *Wear* **1977**, *43*, 341–349. [CrossRef]

23. Lai, X.; Li, H.; Li, C.; Lin, Z.; Ni, J. Modelling and analysis of micro scale milling considering size effect, micro cutter edge radius and minimum chip thickness. *J. Mater. Process. Technol.* **2008**, *48*, 1–14. [CrossRef]

24. Coelho, R.T.; Diniz, A.E.; da Silva, T.M. An Experimental Method to Determine the Minimum Uncut Chip Thickness (h min) in Orthogonal Cutting. *Procedia Manuf.* **2017**, *10*, 194–207. [CrossRef]

25. Harasymowicz, J.; Gawlik, J.; Warziniak, W. Sposób określania minimalnej grubości warstwy skrawanej według patentu P. 252621. *Politech. Krak.* **1900**.

26. Krajewska-Śpiewak, J.; Gawlik, J. A Method for Determination of the Minimal Thickness of the Cutting Layer during Precision Machining Performed with the Indexable Cutting Tools. *SSP* **2017**, *261*, 50–57. [CrossRef]

27. Gawlik, J.; Krajewska-Śpiewak, J.; Zębala, W. Identification of the Minimal Thickness of Cutting Layer Based on the Acoustic Emission Signal. *KEM* **2016**, *686*, 39–44. [CrossRef]

Publisher's Note: MDPI stays neutral with regard to jurisdictional claims in published maps and institutional affiliations.

 *micromachines*

Article

Experimental Investigation on Laser Assisted Diamond Turning of Binderless Tungsten Carbide by In-Process Heating

Kaiyuan You [1], Fengzhou Fang [1,2,*], Guangpeng Yan [1] and Yue Zhang [1]

1   Centre of Micro/Nano Manufacturing Technology (MNMT),
    State Key Laboratory of Precision Measuring Technology & Instruments, Tianjin University,
    Tianjin 300072, China; youkaiyuan@tju.edu.cn (K.Y.); gpyan@tju.edu.cn (G.Y.); yuede1993@163.com (Y.Z.)
2   Centre of Micro/Nano Manufacturing Technology (MNMT-Dublin),
    School of Mechanical & Materials Engineering, University College Dublin, Dublin 4, Ireland
*   Correspondence: fzfang@tju.edu.cn

Received: 26 November 2020; Accepted: 11 December 2020; Published: 14 December 2020

Abstract: Binderless tungsten carbide (WC) finds widespread applications in precision glass molding (PGM). Grinding and polishing are the main processes to realize optical surface finish on binderless WC mold inserts. The laser assisted turning (LAT) by in-process heating is an efficient method to enhance the machinability of hard and brittle materials. In this paper, laser heating temperature was pre-calculated by the finite element analysis, and was utilized to facilitate laser power selection. The effects of rake angle, depth of cut, feed rate, and laser power are studied experimentally using the Taguchi method. The variance, range, and signal-to-noise ratio analysis methods are employed to evaluate the effects of the factors on the surface roughness. Based on the self-developed LAT system, binderless WC mold inserts with mirror finished surfaces are machined using the optimal parameters. PGM experiments of molding glass lenses for practical application are conducted to verify the machined mold inserts quality. The experiment results indicate that both the mold inserts and molded lenses with the required quality are achieved.

Keywords: laser assisted turning; tungsten carbide; diamond turning; finite element analysis

1. Introduction

The binderless tungsten carbide (WC), as the typical hard and brittle material, has been widely applied in the field of optics, photonics and life science owing to its ideal physical properties [1]. However, the ultra-high hardness and brittleness of the WC material also decrease the machinability and greatly limit the feasible machining approaches. At present, mirror finished surfaces of binderless WC mainly relying on grinding and polishing, which is time consuming and infeasible for large-sag concave surface shape [2]. There is a bottleneck of severe tool wear in using single point diamond turning (SPDT) [3]. Too small of a critical ductile-to-brittle transformation (DBT) thickness of hard and brittle materials [4] also brings high cost and labile turning surface integrity [5]. Thus, the more reliable and efficient approach of machining binderless WC is urgent [6,7].

On account of the materials' improved machinability at elevated temperature, laser assisted turning (LAT) is reliable for hard and brittle material machining and is getting more applications. With this approach, the machining efficiency, cutting force, and tool wear can be greatly improved. At present, there are two primary kinds of laser assisted turning forms, including pre-heat laser assisted turning (Pre-LAT) and in-process-heat laser assisted turning (In-LAT) [8]. The Pre-LAT method locates the laser spot above the cutter and pre-heat the cutting region material ahead of tool interaction.

Pre-LAT has a mature development and is widely applied in the traditional cylindrical turning [9] but restricted in the field of end-face ultra-precision diamond turning due to the structural constraint. On the other hand, the In-LAT method guides the laser beam passes through the transparent tool to heat material locally, thereby decreasing the hardness and altering the fracture toughness as soon as the tool interacts with the material [10]. The In-LAT technology brings better machinability, makes the ductile turning of hard and brittle materials possible, and has been a consummate choice for hard and brittle material ultra-precision machining.

In the past decade, there have been an increasing number of studies carried out on the In-LAT of hard and brittle materials. Langan et al. [11] study the laser parameter effects of In-LAT machining on residual stress and phase purity of monocrystalline silicon. Chen et al. [12] also analyze monocrystalline silicon subsurface damage and phase transformation caused by In-LAT through tapper cutting experiment and molecular dynamic simulation. Langan et al. [13] prove the in-process laser heating can reduce the residual stress of sapphire surface finishing and further demonstrating the applicability of the In-LAT method on nominally transparent brittle materials. Park et al. [14] discuss the surface finish and cutting force improvement of bulk metallic glass using sapphire tool with the assistance of laser in-process heat and textured rake face. Di et al. [15] performed In-LAT experiments, proving the In-LAT method can promote the ductile mode material removal of WC, but without lucubrated analysis and discussion. So far, little study discusses the In-LAT parameter effects on the surface quality of binderless WC.

Finite element analysis (FEA) is a superior method to calculate the thermal field based on the numerical calculation [16]. Since the measurement resolution of commercial thermal imaging cameras is restricted by the diffraction limit, the peak temperature during laser heating and nanoscale machining process cannot be measured precisely, but can be calculated using the FEA method. Kamlesh et al. [16] calculate the silicon temperature distribution under Gaussian profile laser beam heating regarding the material high-pressure phase change. Li et al. [17] compare the temperature field of Al_2O_3 ceramic material with volumetric and surface laser heating sources. Shang et al. [18] utilize the three-dimensional transient heat conduction model to predict the temperature distribution caused by a freeform laser trajectory, whose results inversely facilitate the laser parameter selections in the laser assisted milling. At present, there is no study has discussed the temperature field of binderless WC under spiral moving laser heating, which can guide the In-LAT laser parameters selection.

Various parameters of In-LAT have a significant influence on the binderless WC mold inserts finish quality, but to our best knowledge, there is little comprehensive discussion until now. In this study, the laser power heating temperature was analyzed using both the numerical calculation and FEA, acquiring the approximate range of optimal laser power. Furthermore, the experiment design methodology, the Taguchi design, was utilized to explore the optimal In-LAT parameters for binderless WC. The effects of tool rake angle, machining depth of cut (DoC), feed rate, and laser power were studied experimentally. The analysis of variance (ANOVA), signal to noise (S/N) ratio, and range analysis methods were utilized to select the optimal parametric combination and verify the validity of the experiment. With the optimal parameters, binderless WC mold inserts were machined based on the self-developed LAT system, and were used in the precision glass molding (PGM) experiment subsequently. Both the mold inserts and molded lenses with ideal surface quality were achieved, verifying the In-LAT machined mold insert can be successfully applied for the replicative mass production of glass lenses with mirror finished surface.

The rest of this paper is organized as follows. Section 2 calculates the workpiece temperature under various laser power heating. Section 3 introduces the experimental setup and the parameter design. The analysis of the Taguchi experiment result is described in Section 4.1. The final mold inserts machined by the In-LAT method and the PGM experiment are presented in Section 4.2. Finally, the major conclusions and the following works are presented in Section 5.

2. Thermal Field Analysis of Laser Heating

The material's temperature is a vital index for In-LAT and has a considerable impact on the machining quality. The appropriate heating temperature could decrease the material's hardness and increase the machinability, but the excess temperature will introduce undesired thermal damage and severe diamond tool wear. It is essential to maintain the laser heating temperature within the optimal range, which is mainly dependent on the irradiated laser power. The numerical calculation could facilitate the thermal field analysis and economize the experimental works. Considering the thermal initial condition and complex boundary conditions, the workpiece thermal field can be precisely calculated. Before the laser heating, the initial thermal condition of the WC workpiece can be determined by the uniform temperature T_0 as

$$T(0,t) = T_0. \tag{1}$$

There are many boundary conditions during the laser heating process, which can be divided into the laser heating effect, thermal radiation [19], and cutting fluid convection [20]. For the Gauss laser spot on the $z = 0$ plane, the heat flux of spiral trajectory moving laser spot can be described as

$$q_l(x,y,0,t) = \frac{2P\delta}{\pi r^2} \exp\left\{ \frac{2\left\{[x - (x_0 - ft)\cos(\pi nt/30)]^2 + [y - (x_0 - ft)\sin(\pi nt/30)]^2\right\}}{r^2} \right\}, \tag{2}$$

where P (W) refers to the laser power, r (μm) is the laser spot radius, δ is the laser absorptivity, and the laser scanning path can be determined by the initial position $(x_0,0)$, feed rate f (μm/rev), and rotation speed n (rev/min). In addition, the cutting fluid convection can be calculated by

$$q_c(x,y,z,t) = h(T_f - T), \tag{3}$$

where h (W·m^{-2}·K^{-1}) represents the convective heat transfer coefficient and should be measured by the experiment, T_f refers to the cutting fluid temperature. Thermal radiation is

$$q_r(x,y,z,t) = \sigma\varepsilon(T_{amb}^4 - T^4), \tag{4}$$

where the $\sigma = 5.67 \times 10^{-8}$ W/(m^2·K^4) is the Stefan-Boltzmann constant, and the ε represents the surface emissivity, T_{amb} means the ambient temperature. The basic equation for the analysis of heat conduction is the Fourier's law

$$q_n = -K_n \frac{\partial T}{\partial n}, \tag{5}$$

where heat flux q_n is the heat transfer rate in the n direction per unit area perpendicular to the direction of heat flow. $\partial T/\partial n$ (K/m) is the temperature gradient in the direction n. The K_n (W·m^{-1}·K^{-1}) is the thermal conductivity in n direction. However, the workpiece material is thought to be isotropic in this study, indicating that the K_n is a constant value. Owing to there being no volumetric energy addition, so the three-dimensional thermal conduction within the workpiece can be calculated by

$$\frac{\partial T}{\partial t} = k\nabla^2 T, \tag{6}$$

where k is the thermal diffusion coefficient in m^2/s, satisfying $k = K/(\rho \cdot C_p)$. K refers to the material thermal conductivity coefficient, ρ represents the density of the material (kg/m^3), and C_p is the specific heat (J/kg·K) of the material.

FEA is a powerful numerical calculation strategy, and was utilized to compute the temperature field as shown in Figure 1. The laser heating WC workpiece model was established with the assistance of COMSOL software. The triangular mesh type was adopted and firstly generated in the workpiece upper surface and then sweep in the thickness direction. In particular, the FEA mesh characteristic size was controlled within 1 μm to 40 μm to avoid the tedious calculation time and severe calculation distortion. The workpiece coordinate system was established with the origin coincident with the workpiece center. The WC workpiece is regarded as isotropic with 4 mm diameter and 0.5 mm thickness. Furthermore, the 1064 nm continuous wave laser with a radius of 85 μm is loaded on the workpiece upper surface ($z = 0$). The laser energy distribution conforms to the Gauss theorem. The relative motion between laser spot and workpiece in the FEA model is consistent with the machining scene. The laser spot feeds along -x direction with 1 mm/min speed while the workpiece rotates with the constant 2000 rev/min. The typical thermodynamic parameters such as laser absorptivity refer to the pre-researchers' work, which ensures the accuracy and reliability of the simulation results. The initial workpiece temperature is assumed to be 293 K. For clarity, the FEA simulation parameters have been summarized in Table 1.

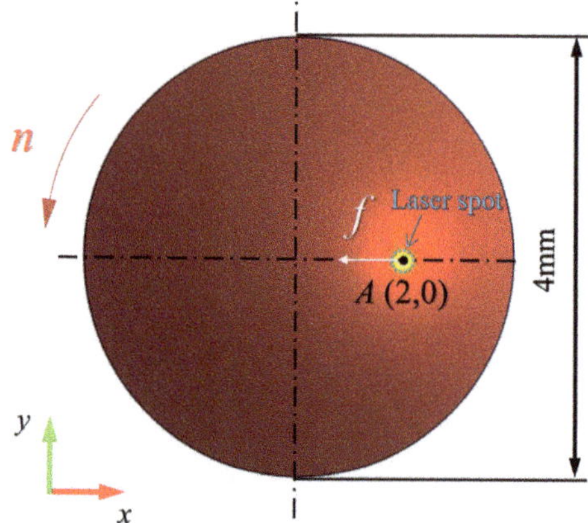

Figure 1. Finite element analysis (FEA) model of laser heating

Table 1. Finite element analysis (FEA) simulation parameters.

Parameters	Value
Laser wavelength (λ)	1064 nm
Laser power (P)	5 W, 10 W, 15 W, 20 W
Laser spot radius (r)	85 μm
Laser translational velocity (f)	1 mm/min
Workpiece material	Tungsten carbide
Rotation speed (n)	2000 rev/min
Workpiece diameter	4 mm
Workpiece thickness	0.5 mm
Initial temperature (T_0)	293 K
Emissivity (ε)	0.75 [19]
Absorptivity (δ)	0.23 [21]
Convective heat transfer coefficient (h)	300 W/(m^2·K) [20]

Since the laser spot center is coincident with the diamond tool tip and the radius of laser spot (85 μm) is much larger than the feed distance per cycle (0.5 μm), the laser heat accumulation effect should be considered. The laser heating process should be computed when the laser spot covers the specific machining point. For the specifical point $A(2,0)$, the workpiece was heated by the moving laser beam from the initial position (2.085,0) of beam center to the end position (1.915,0) with the 1mm/min velocity. The FEA model calculates the thermal field of the laser heating with four power levels (5 W, 10 W, 15 W, 20 W) based on Equations (1)–(6). The workpiece highest temperature, and the temperature history of specific point A have been recorded and plotted in Figure 2. When the laser beam center moves to the A point, the temperature of point A is same as the highest temperature, which increase to 554.0 K, 802.2 K, 1044.7 K, and 1294.9 K for 5 W, 10 W, 15 W, and 20 W laser power respectively. If the workpiece temperature is much higher than the diamond graphitization temperature 1000 K [22], it will result in severe tool wear and thermal damage. Thus, higher laser power will not be considered. The parameters of 5 W, 10 W, 15 W, 20 W were adopted in the following orthogonal experiment.

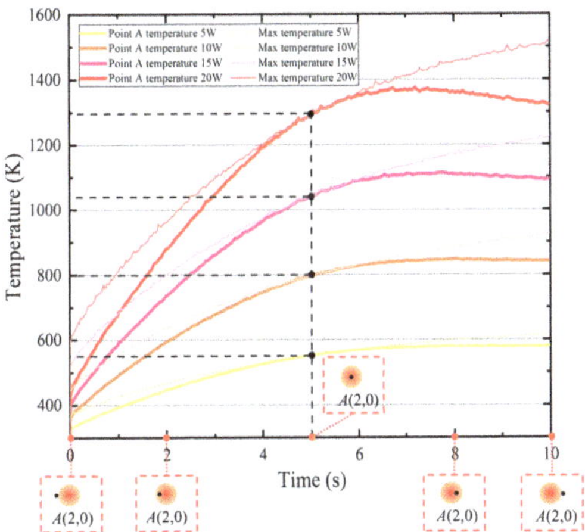

Figure 2. Historical temperature information of binderless tungsten carbide (WC) workpiece under different laser power heating.

3. Experimental Approach

3.1. In-Process-Heat Laser Assisted Turning (In-LAT) Experiment Setup

The experiment was carried on based on an ultra-precision three-axis lathe with the self-developed LAT system as shown in Figure 3a. The x-axis and z-axis of lathe drive the spindle and diamond tool moves linearly in x and z directions, respectively. The binderless WC was mounted on the aluminum substrate and vacuum-chucked onto the lathe spindle. The workpiece has a concave spherical surface with 7.5 mm diameter and 7 mm curvature, and rotates with the lathe spindle, realizing the cutting movement and material removal.

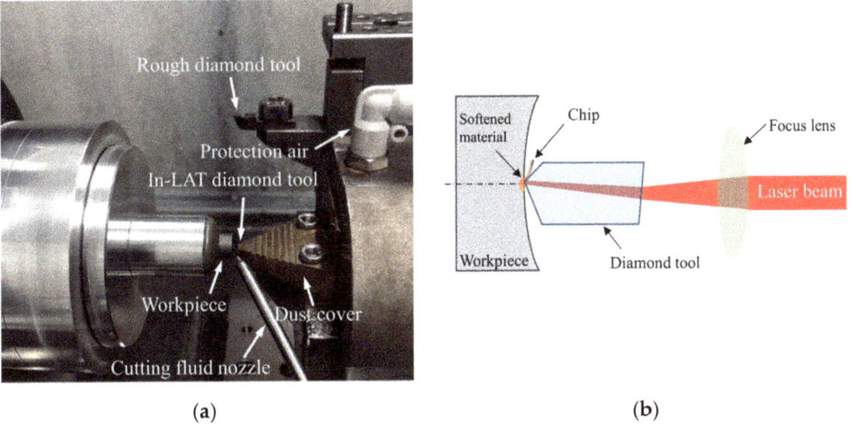

(a)   (b)

Figure 3. (a) The experimental setup, and (b) the schematic diagram of the in-process-heat laser assisted turning (In-LAT) system.

The self-developed LAT system was set up on the z-guide platform of lathe without deflection. There are several primary parts of the LAT system, including the laser source, optical elements, positioning device, tool holder, and In-LAT diamond tool. Specifically, the Nd: YAG fiber laser can generate continuous wave laser with 1064 nm wavelength, which possesses the Gaussian power distribution (M^2 = 1.126) and can adjust within the range of 2–100 W. The emitted laser is guided into the optical system through the laser fiber, thereby focusing the laser beam into the minimum 14 µm diameter laser spot and can also be enlarged by defocusing. The multi-axis positioning device can drive the optical system to move along three vertical directions realizing the laser spot position alignment. For the purpose of protecting optical lenses, the cutting chips are separated from the optical system via the protection air and dust cover. Moreover, since the natural diamond has an inherent variability in mechanical and optical properties, the chemical vapor deposition (CVD) monocrystalline diamond with extremely uniform properties was chosen to fabricate the In-LAT diamond tool.

3.2. Parameters Design

Taguchi method is an experimental design methodology to study the effects of multi-factors and multi-levels and was used for investigating the In-LAT optimal laser power and machining parameters of binderless WC in this study. The experimental parameters were designed referring to orthogonal array, whose factors are independent of each other and can be evaluated separately. The effect of one individual factor would not affect the estimation of other factors. Taguchi method can be employed to greatly reduce experiments by selecting some representative conditions from the comprehensive experiment according to the orthogonality.

There are various factors that have a significant influence on the surface finish quality, including the machining parameters and laser parameters. The effects of laser power, tool rake angle, DoC, and feed rate are considered with 5 levels herein as listed in Table 2. The level values of machining factors are determined by the preliminary experiments. The laser powers are selected by the numerical calculation results. In particular, the blank group was designed for the variance analysis and separating the impact of the accidental error. Finally, the $L_{25}(5^5)$ standard orthogonal array was selected and utilized to perform experiments.

Table 2. Orthogonal parameters and corresponding values

Factors	Units	Levels of Factors				
		Level 1	Level 2	Level 3	Level 4	Level 5
Rake angle	°	0	−15	−25	−35	−45
Depth of cut	µm	1	2	4	6	8
Feed rate	µm/rev	0.5	1	2	3	4
Laser power	W	0	5	10	15	20
Blank group		1	2	3	4	5

In order to ensure the rationality of the orthogonal experiment, the rest machining parameters remain constant as listed in Table 3. Experiments were conducted on the same workpiece, which rotates with a spindle speed of 2000 rpm. The diamond tools with a nominal 10° clearance angle and 0.3 mm nose radius were used. Each orthogonal experiment ensured that the cutting edges of the In-LAT diamond tools are in good condition and have the homogeneous quality. Moreover, an 85 µm radius laser spot was employed for enough laser beam cover angle on the cutting edge, and was obtained with the 1.96 mm defocusing length. The laser spot position has been accurately adjusted where the laser spot center is coincident with the diamond tool tip. For facilitating the visual analysis, the laser beam path at the ideal position was modeled by the optical simulation. Almost the half laser beam is reflected on the tool rake face owing to the total reflection principle, the emitted laser spot pattern is plotted in Figure 4a with semi-Gauss energy distribution. The reflection part of laser beam continuously heats the tool holder as shown in Figure 4b. All the experiments were carried on using the same cutting fluid

(ISOPAR-H) assistance in the same fluid spraying direction. The deep thermal damage layer caused by high-power laser was pre-removed by the rough diamond tool. The workpiece surfaces before turning were polished using the diamond paste (grain size 0.5 µm) before each orthogonal experiment, guaranteeing the consistency of workpiece surface quality.

Table 3. Constant machining parameters

Parameters	Description
Workpiece material	Binderless tungsten carbide
Rotation speed	2000 rpm
Tool material	CVD monocrystalline diamond
Clearance angle	10°
Nose radius	0.3 mm
Laser spot radius	85 µm
Laser position	Beam center coincides with tool tip
Cutting fluid	ISOPAR-H

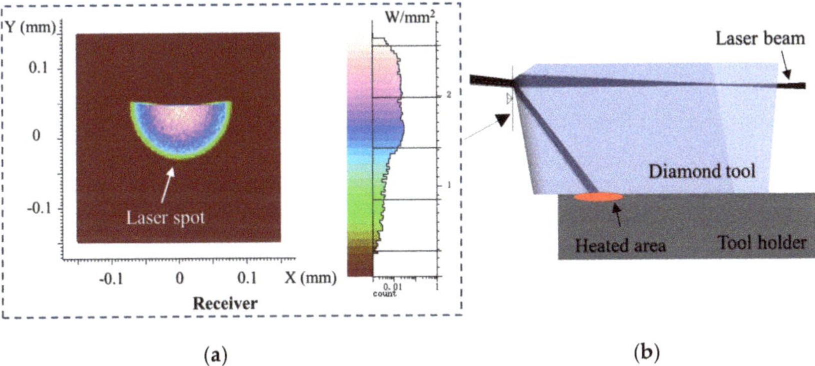

Figure 4. Optical simulation results. (**a**) Emitted laser spot from diamond tool and (**b**) laser beam path in the In-LAT diamond tool.

4. Results and Discussion

4.1. Effect of Test Parameters on Surface Roughness

In this study, the surface roughness was chosen as the evaluation index of surface finish quality. The surface roughness of the machined workpiece was measured by a laser confocal microscope (OLYMPUS LEXT OLS 4000, Olympus, Tokyo, Japan) with 20× objective lenses and 4× digital zoom. The scanning field is set as (161 × 161 µm²). All measurement data have been disposed with the 80 µm high-pass filter to remove the intermediate frequency signal introduced by the surface form. Surfaces machined by each parameter group were measured three times to eliminate the effect of the accidental error. The mean roughness has been utilized to analyze the orthogonal experiment factors' impact on the surface finish quality.

4.1.1. Analysis of Variance

The ANOVA method can be used to distinguish the difference between the test results caused by the factor variation from the impact of the accident error, and can give a reliable quantitative estimate. Based on the measurement results, ANOVA was applied to study the significance of the input factors on the surface roughness at a 95% confidence level. The factors' degree-of-freedom, sum-of-squares, mean-of-squares, *F*-value, *p*-value, and contribution rate are calculated and listed in Table 4.

Table 4. Analysis of variance (ANOVA) results from mean surface roughness

Factors	Degree-of-Freedom	Sum-of-Squares	Mean-of-Squares	F-Value	p-Value	Contribution Rate
Rake angle	4	7491.8	1872.9	5.36	0.021	48.85%
DoC	4	1597.8	399.4	1.14	0.402	10.42%
Feed rate	4	1238.6	309.6	0.89	0.514	8.08%
Laser power	4	2211.8	552.9	1.58	0.269	14.42%
Error	8	2795.5	349.4	-	-	18.23%
Total	24	15335.4	-	-	-	-

The p-value reflects the significance of orthogonal factors. The factor with a smaller p-value indicates it has a more significant influence on the experimental results. The p-values of rake angle, DoC, feed rate, and laser power are 0.021, 0.402, 0.514, and 0.269 respectively. In particular, the ANOVA results declare that the blank group has a larger influence than the DoC and feed rate, indicating their effect is not significant within the designed parameter range. The contribution rate of rake angle and laser power counts 48.85% and 14.42%, respectively. It is evident that the rake angle and laser power are the main factors affecting the surface roughness during the binderless WC machining, and that the rake angel of diamond tool is the most influential factor.

4.1.2. Analysis of the Signal-to-Noise Ratio and Mean Value

The S/N ratio is an important index of the Taguchi design robustness. A reasonable parameter combination with a minimize noise factor effect can be selected through the results analysis, thereby ensuring the stability of the machining process. If the value of the signal-to-noise ratio (S/N) is larger, the effect of the experimental noise factor under this parameter value is smaller. In this study, the surface roughness is chosen as the evaluated index and desired to be small. Thus, the principle of the smaller-the-better was adopted in the S/N ratio analysis. The effects of individual factors on the surface roughness characteristics were analyzed as listed in Table 5. The analysis results indicate that the parameter combination of −25° rake angle, 6 μm DoC, 1 μm/rev feed rate, and 10 W laser has minimal effect of noise factor as shown in Figure 5. The delta of the blank group presents the smallest value (3.12 dB), verifying the validity of the experiment design.

Table 5. Response table for mean surface roughness

Control Factors	Resultant Surface Roughness						
	Level 1	Level 2	Level 3	Level 4	Level 5	Delta	Row Rank
Mean of S/N ratio (dB)							
Rake angle	−26.22	−28.08	−24.86	−29.88	−36.15	11.30	1
DoC	−28.26	−29.01	−31.98	−26.70	−29.23	5.28	2
Feed rate	−28.86	−25.75	−30.08	−30.36	−30.14	4.61	3
Laser power	−28.76	−27.77	−27.04	−30.05	−31.56	4.52	4
Blank group	−30.48	−29.63	−28.26	−29.45	−27.36	3.12	5
Mean of surface roughness (nm)							
Rake angle	21.00	28.40	18.80	39.80	66.20	47.40	1
DoC	31.20	30.40	48.80	25.80	38.00	23.00	3
Feed rate	34.80	22.80	33.60	44.00	39.00	21.20	4
Laser power	33.00	27.20	25.40	36.80	51.80	26.40	2
Blank group	46.4	33.20	27.80	32.80	34.00	18.60	5

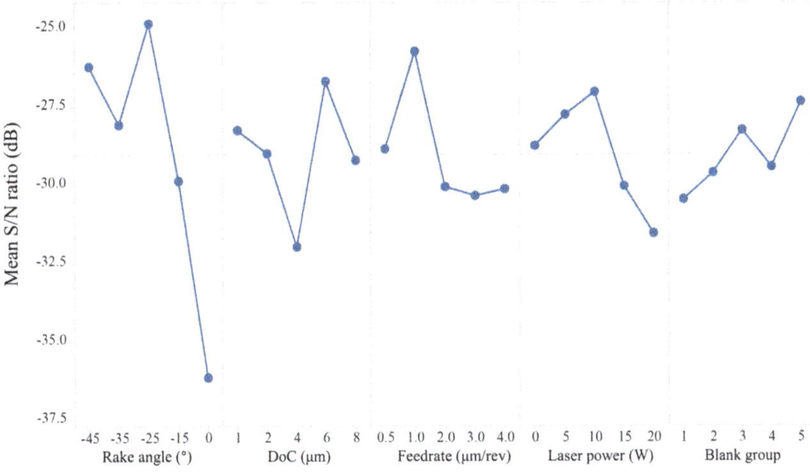

Figure 5. Main effect of mean S/N ratios corresponding to the surface roughness.

In order to optimize the experimental parameters, the statistical results of factor impact have been analyzed. The mean value and delta of surface roughness for individual factors at the same level have been calculated as summarized in Table 5. In particular, the delta of the blank group has the biggest value, verifying the experiment validity again. The experimental results declare the −25° tool rake angle has the minimum surface roughness as shown in Figure 6a. The negative rake angle introduces large hydrostatic stress, which is beneficial for the high-pressure phase transformation of hard and brittle materials [23] and has much better results than zero-rake angle tools. However, the large negative rank angle tool can also bring the difficulty of chip removal and large cutting force which also result in the deteriorating machining quality. Besides, the 6 μm DoC has a lower surface roughness as shown in Figure 6b. However, the DoC effect of the experiment has obvious randomicity, indicating the DoC has a slight influence on the surface roughness within the given level range. On the other hand, 1.0 μm/rev is thought to be the optimal feed rate of binderless WC machining as shown in Figure 6c. The small feed rate makes the undeformed chip thickness (UCT) smaller than the critical DBT depth of WC. However, the smallest feed rate level (0.5 μm/rev) is so small that resulted in UCT is likely to smaller than the tool edge radius and brings a large effective negative rake angle [24]. Moreover, the optimal surface quality was obtained with 10 W laser power as shown in Figure 6d. There is obvious thermal damage and deteriorate machining quality under 20 W laser power as shown in Figure 7a. Furthermore, since the laser heating locally, there are some stuck high-temp chips on the machined WC surface. The stuck chip can be removed by the polishing process but not work for wipe as shown in Figure 7b. From this discussion, it can be concluded that the optimal parametric combination for minimum surface roughness was the group of the −25° tool rake angle, 6 μm DoC, 1.0 μm/rev feed rate, and 10 W laser power, which is consistent with the result of S/N ratio analyze.

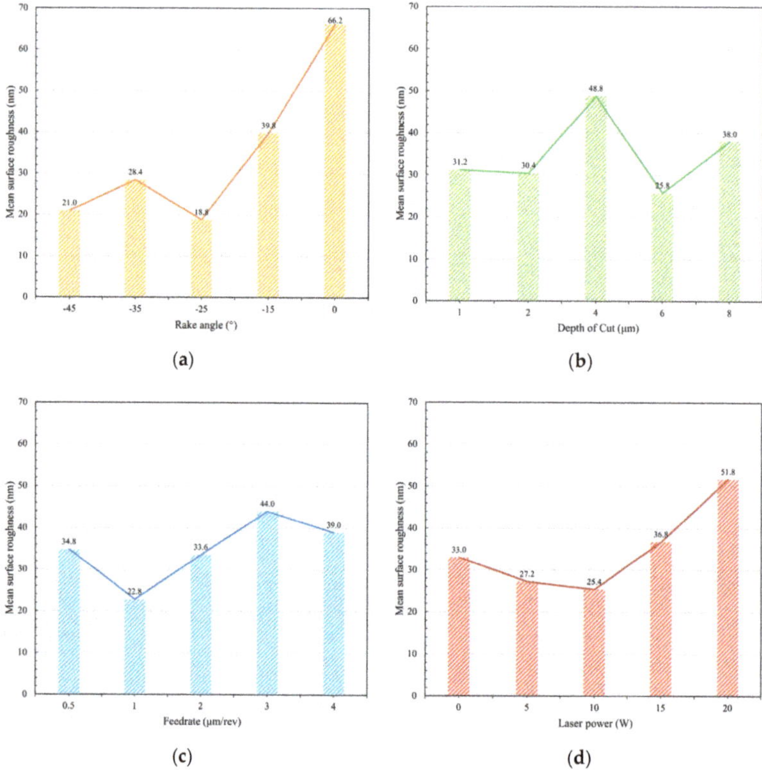

(a)

(b)

(c)

(d)

Figure 6. Surface roughness correlation with the variation of (**a**) rake angle, (**b**) DoC, (**c**) feed rate, and (**d**) laser power.

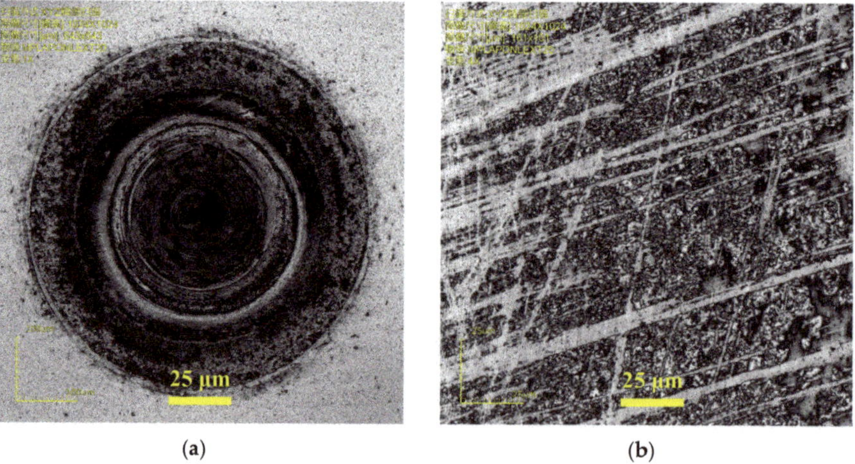

(a)

(b)

Figure 7. Laser microscope images of (**a**) thermal damage area under 20 W laser; (**b**) high-temp chip stuck on the machined surface.

4.2. Mold Insert and Molded Glass Quality

With the obtained optimal parameters, two concave mold inserts with ideal surface quality have been machined on binderless WC via the self-developed LAT system as shown in Figure 8a. The inserts possess a 1.62 mm diameter concave aspherical surface and 0.4 mm width platform. Two inserts were machined using the same diamond tool seven times to verify the parametric stability. In order to verify the mold inserts' quality, the PGM experiment was performed on an aspherical lens molding machine (DTK-LMR-3300 V2, Daehoteck, Changwon-si, Korea). The glass selected for the molding experiment was a moldable glass (D-ZLAF52LA, CDGM, Chengdu, China), whose transition temperature is 546 °C. Molded lenses were produced under 0.3 Mpa pressure and 586 °C temperature as shown in Figure 9a.

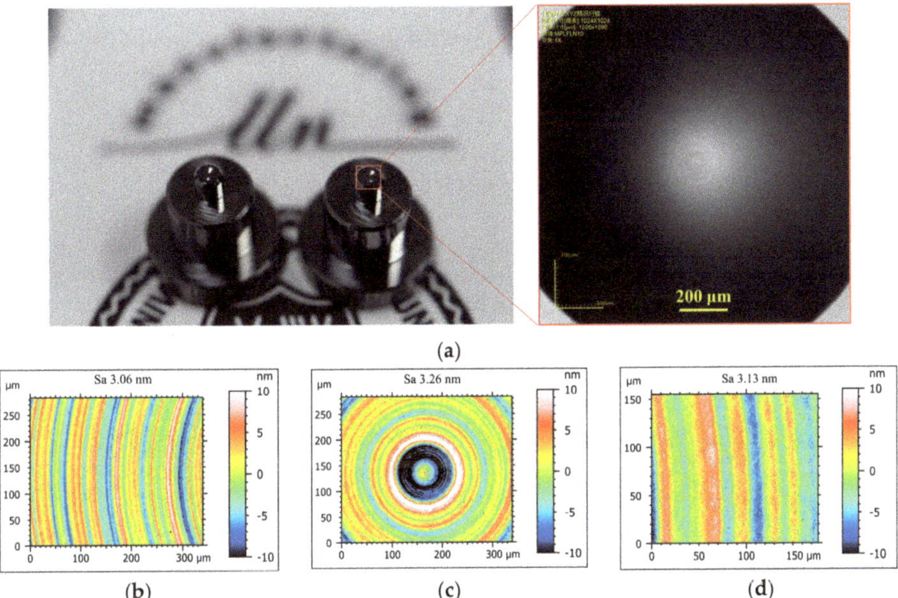

Figure 8. In-LAT machined mold inserts, (**a**) the measured mold surface morphology, (**b**) the fringe area, (**c**) central area, and (**d**) outer platform area.

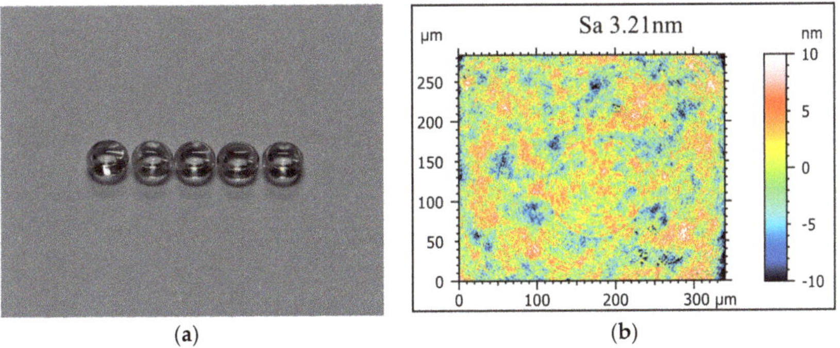

Figure 9. Molded lenses (**a**) the molded lenses, (**b**) the measured surface morphology of the central region.

4.2.1. Surface Finish

Both the mold inserts and molded lenses with homogeneous quality are achieved, and the surface morphology was measured by a white light interferometer (S Neox, Sensofar, Barcelona, Spain). The 20× objective lenses were used under the VSI measurement mode for the moderate scanning field (0.340 × 0.283 μm²). The measurement data has also been disposed with the 80 μm high-pass filter to remove the intermediate frequency signal introduced by the surface form. The edge part is clipped to suppress the edge effect of filtering. Furthermore, the fringe area of mold inserts possesses the surface roughness of *Sa* 3.06 nm as shown in Figure 8b. The center area presents larger surface roughness *Sa* 3.26 nm owing to the tool setting error and slow cutting speed as shown in Figure 8c. Moreover, the smaller field (0.176 × 0.154 μm²) has been measured with *Sa* 3.13 nm as shown in Figure 8d on the platform of mold insert owing to the narrow width of the platform. The molded glass optics have homogeneous quality with the mold insert as shown in Figure 9b, with the ideal surface quality of *Sa* 3.21 nm.

4.2.2. Form Error

The form error of mold inserts and molded lenses were measured using a form measurement instrument (UA3P-300, Panasonic, Osaka, Japan). The diamond probe with a radius of 2 μm was used to scan the optics at a scanning speed of 0.2 mm/s. Since almost half part of the laser beam total reflect on the tool rake face and absorbed by the tool holder as shown in Figure 4, the tool holder is heated continuously during the In-LAT machining process, thereby introducing inevitable thermal expansion, and then resulted in poor alignment accuracy. The introduced tool setting error makes the form error of mold insert is not ideal with PV 0.757 μm as shown in Figure 10. The molded lenses also present a complementary form error pattern with PV 0.814 μm, which is unqualified for the practical application.

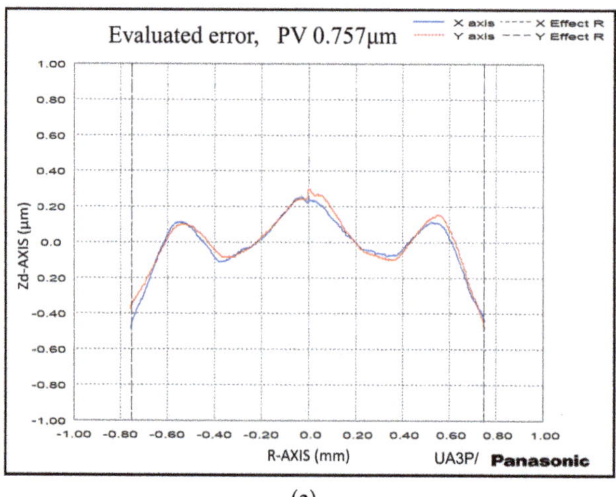

(a)

Figure 10. *Cont.*

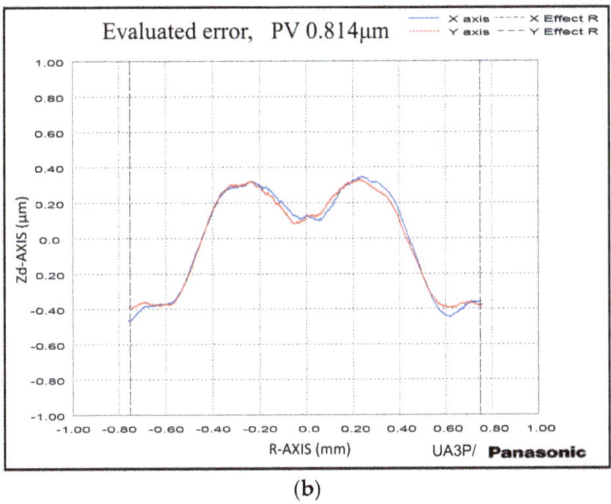

(**b**)

Figure 10. The form error of (**a**) the mold inserts and (**b**) molding lenses.

5. Conclusions

The laser power heating temperature was analyzed with the numerical calculation and FEA model firstly, acquiring the approximate range of optimal laser power. The orthogonal In-LAT experiment with the statistical analysis method of AVONA, S/N ratio, and range analysis was conducted to study the effect of tool rake angle, machining DoC, feed rate, and laser power. The optimal parametric combination for minimum surface roughness and maximal S/N ratio was achieved using the −25° tool rake angle, 6 μm DoC, 1.0 μm/rev feed rate, and 10 W laser power. With the optimal parameters, the mold inserts with ideal surface quality were machined on the binderless WC based on the self-developed In-LAT system. The PGM experiment was performed subsequently to verify the mold inserts' quality. The experiment results indicate that surface roughness of 3.26 nm and 3.21 nm in Ra have been achieved on the central area of inserts and lenses respectively, verifying the superiority and robustness of the optimal parametric combination. The form error of insert and lenses with PV 0.757 μm and PV 0.814 μm were obtained respectively, which has room for further improvement in the following investigation.

Author Contributions: Conceptualization, investigation, methodology, data curation, writing—original draft preparation, K.Y.; conceptualization, methodology, writing—review and editing, supervision, project administration, funding acquisition, F.F.; validation and formal analysis, G.Y.; validation and formal analysis, Y.Z. All authors have read and agreed to the published version of the manuscript.

Funding: This research was funded by the National Key Research & Development Program (no. 2016YFB1102203), the National Natural Science Foundation of China (no. 61635008), and the '111' project conducted by the State Administration of Foreign Experts Affairs and the Ministry of Education of China (no. B07014).

Acknowledgments: The authors would like to express their sincere thanks to Yongxu Xiang, Jiaming Dong, and Zhen Li for their assistance in the preparation of the experiments.

Conflicts of Interest: The authors declare no conflict of interest. The funders had no role in the design of the study; in the collection, analyses, or interpretation of data; in the writing of the manuscript, or in the decision to publish the results.

References

1. Roeder, M.; Guenther, T.; Zimmermann, A. Review on Fabrication Technologies for Optical Mold Inserts. *Micromachines* **2019**, *10*, 233. [CrossRef] [PubMed]
2. Yan, G.P.; Fang, F.Z. Fabrication of optical freeform molds using slow tool servo with wheel normal grinding. *Cirp Ann.* **2019**, *68*, 341–344. [CrossRef]
3. Fähnle, O.; Doetz, M.; Dambon, O.; Klocke, F.; Vogt, C.; Rascher, R. Ductile mode single point diamond turning (SPDT) of binderless tungsten carbide molds. In Proceedings of the Optical Manufacturing and Testing XII, San Diego, CA, USA, 20–22 August 2018. [CrossRef]
4. Doetz, M.; Dambon, O.; Klocke, F.; Fähnle, O. Influence of coolant on ductile mode processing of binderless nanocrystalline tungsten carbide through ultraprecision diamond turning. In Proceedings of the SPIE Optical Engineering + Applications, San Diego, CA, USA, 9–13 August 2015. [CrossRef]
5. Wang, J.S.; Fang, F.Z.; Yan, G.P.; Guo, Y. Study on diamond cutting of ion implanted tungsten carbide with and without ultrasonic vibration. *Nanomanuf. Metrol.* **2019**, *2*, 177–185. [CrossRef]
6. Fang, F.Z. Atomic and close-to-atomic scale manufacturing: Perspectives and measures. *Int. J. Extrem. Manuf.* **2020**, *2*, 030201. [CrossRef]
7. Mathew, P.T.; Rodriguez, B.J.; Fang, F.Z. Atomic and Close-to-Atomic Scale Manufacturing: A Review on Atomic Layer Removal Methods Using Atomic Force Microscopy. *Nanomanuf. Metrol.* **2020**, *16*, 1–20. [CrossRef]
8. You, K.Y.; Yan, G.P.; Luo, X.C.; Gilchrist, M.D.; Fang, F.Z. Advances in laser assisted machining of hard and brittle materials. *J. Manuf. Process.* **2020**, *58*, 677–692. [CrossRef]
9. Song, H.W.; Dan, J.Q.; Li, J.L.; Du, J.; Xiao, J.F.; Xu, J.F. Experimental study on the cutting force during laser-assisted machining of fused silica based on the Taguchi method and response surface methodology. *J. Manuf. Process.* **2019**, *38*, 9–20. [CrossRef]
10. Ravindra, D.; Ghantasala, M.K.; Patten, J. Ductile mode material removal and high-pressure phase transformation in silicon during micro-laser assisted machining. *Precis. Eng.* **2012**, *36*, 364–367. [CrossRef]
11. Langan, S.M.; Ravindra, D.; Mann, A.B. Process parameter effects on residual stress and phase purity after microlaser-assisted machining of silicon. *Mater. Manuf. Process.* **2018**, *33*, 1578–1586. [CrossRef]
12. Chen, X.; Liu, C.L.; Ke, J.Y.; Zhang, J.G.; Shu, X.W.; Xu, J.F. Subsurface damage and phase transformation in laser-assisted nanometric cutting of single crystal silicon. *Mater. Des.* **2020**, *190*, 108524. [CrossRef]
13. Langan, S.M.; Ravindra, D.; Mann, A.B. Mitigation of damage during surface finishing of sapphire using laser-assisted machining. *Precis. Eng.* **2019**, *56*, 1–7. [CrossRef]
14. Park, S.S.; Wei, Y.; Jin, X.L. Direct laser assisted machining with a sapphire tool for bulk metallic glass. *CIRP Ann.* **2018**, *67*, 193–196. [CrossRef]
15. Kang, D.; Navare, J.; Su, Y.; Zaytsev, D.; Shahinian, H. Observations on Ductile Laser Assisted Diamond Turning of Tungsten Carbide. In Proceedings of the Freeform Optics, Washington, DC, USA, 10–12 June 2019. [CrossRef]
16. Suthar, K.J.; Patten, J.; Dong, L.; Abdel-Aal, H. Estimation of Temperature Distribution in Silicon During Micro Laser Assisted Machining. In Proceedings of the ASME 2008 International Manufacturing Science and Engineering Conference collocated with the 3rd JSME/ASME International Conference on Materials and Processing, Evanston, IL, USA, 7–10 October 2008. [CrossRef]
17. Li, J.F.; Li, L.; Stott, F.H. Comparison of volumetric and surface heating sources in the modeling of laser melting of ceramic materials. *Int. J. Heat Mass Transf.* **2004**, *47*, 1159–1174. [CrossRef]
18. Shang, Z.D.; Liao, Z.R.; Sarasua, J.A.; Billingham, J.; Dragos, A. On modelling of laser assisted machining: Forward and inverse problems for heat placement control. *Int. J. Mach. Tools Manuf.* **2019**, *138*, 36–50. [CrossRef]
19. Dai, H.F.; Li, S.B.; Chen, G.Y. Comparison of subsurface damages on mono-crystalline silicon between traditional nanoscale machining and laser-assisted nanoscale machining via molecular dynamics simulation. *Nucl. Instrum. Methods Phys. Res. B* **2018**, *414*, 61–67. [CrossRef]
20. Kurgin, S.; Dasch, J.M.; Simon, D.L.; Zou, B.Q. Evaluation of the convective heat transfer coefficient for minimum quantity lubrication (MQL). *Ind. Lubr. Tribol.* **2012**, *64*, 376–386. [CrossRef]
21. Romero, P.A.; Anciaux, G.; Molinari, A.; Molinari, J.F. Insights into the thermo-mechanics of orthogonal nanometric machining. *Comput. Mater. Sci.* **2013**, *72*, 116–126. [CrossRef]

22. Fedoseev, D.V.; Vnukov, S.P.; Bukhovets, V.L.; Anikin, B.A. Surface graphitization of diamond at high temperatures. *Surf. Coat. Technol.* **1986**, *28*, 207–214. [CrossRef]
23. Gilman, J.J. Insulator-metal transitions at microindentations. *J. Mater. Res.* **1992**, *7*, 535–538. [CrossRef]
24. Fang, F.Z.; Wu, H.; Liu, Y.C. Modelling and experimental investigation on nanometric cutting of monocrystalline silicon. *Int. J. Mach. Tools Manuf.* **2005**, *45*, 1681–1686. [CrossRef]

 *micromachines*

Article

Jet Electrochemical Micromachining of Micro-Grooves with Conductive-Masked Porous Cathode

Guochao Fan [1], Xiaolei Chen [1,*] , Krishna Kumar Saxena [2], Jiangwen Liu [1,*] and Zhongning Guo [1]

1   School of Electromechanical Engineering, Guangdong University of Technology, Guangzhou 510016, China; gcfangdut@163.com (G.F.); znguo@gdut.edu.cn (Z.G.)
2   Department of Mechanical Engineering, KU Leuven, 3001 Leuven, Belgium; krishna.saxena@kuleuven.be
*   Correspondence: xlchen@gdut.edu.cn (X.C.); fejwliu@scut.edu.cn (J.L.)

Received: 24 April 2020; Accepted: 29 May 2020; Published: 30 May 2020

Abstract: Surface structures with micro-grooves have been reported to be an effective way for improving the performance of metallic components. Through-mask electrochemical micromachining (TMEMM) is a promising process for fabricating micro-grooves. Due to the isotropic nature of metal dissolution, the dissolution of a workpiece occurs both along the width and depth. Overcut is generated inevitably with increasing depth, which makes it difficult to enhance machining localization. In this paper, a method of electrochemical machining using a conductive masked porous cathode and jet electrolyte supply is proposed to generate micro-grooves with high machining localization. In this configuration, the conductive mask is directly attached to the workpiece, thereby replacing the traditional insulated mask. This helps in achieving a reduction in overcut and an improvement in machining localization. Moreover, a metallic nozzle is introduced to supply a jetted electrolyte in the machining region with enhanced mass transfer via a porous cathode. The simulation and experimental results indicate that as compared with an insulated mask, the use of a conductive mask weakens the electric field intensity on both sides of machining region, which is helpful to reduce overcut and enhance machining localization. The effect of electrolyte pressure is investigated for this process configuration, and it has been observed that high electrolyte pressure enhances the mass transfer and improves the machining quality. In addition, as the pulse duty cycle is decreased, the dimensional standard deviation and roughness of the fabricated micro-groove are improved. The results suggest the feasibility and reliability of the proposed method.

Keywords: micro-groove; electrochemical machining; porous cathode; conductive mask; machining localization; dimensional uniformity

1. Instruction

Together with the development of modern micromanufacturing technologies, the requirements for multiple product function integration and structure miniaturization are becoming more important. Therefore, with an increasing demand of microstructures in industrial applications, there is a recent surge in the research on micro-scale features, which have received extensive attention from both academia and industry [1,2]. As a typical example of surface texture, micro-grooves are widely used in fuel cells, cutting tools, hydrodynamic bearings and heat transfer. For example, micro-grooves prepared on the bipolar plate of proton exchange membrane fuel cells (PEMFC) have a significant impact on the performance and operation efficiency, thereby serving the purpose of increasing the output power density of fuel cells [3,4]. Micro-grooved textures can be fabricated on the surface of cutting tools, which can improve the cutting performance and service life, since it reduces the contact length between chips and cutters, storing lubricant and slowing tool wear [5,6]. Owing to the positive impact on lubrication performance, hydrodynamic bearings textured with micro-grooves have better stability and anti-wear property [7,8].

In recent years, many scholars have conducted extensive research on the efficient fabrication of micro-groove surface structures. The common methods for the fabrication of micro-grooves include micro milling [9], laser machining [10], and micro-electrical discharge machining [11], etc. The application of micro-machining is limited by the hardness of the workpiece, tool strength at diameters < 100 μm, tool wear, and tool vibration, which affects the surface quality. Laser processing can make the workpiece surface prone to severe deterioration layers and micro-cracks, and it is difficult to guarantee the machining accuracy of dimension. Micro-electrical discharge machining suffers from the heat-affected zone around the metallographic structure, electrode wear, and machining cost. Compared with these methods, electrochemical machining (ECM) is more suitable in the micromachining domain for the manufacturing of micro-profiles, thin-walled parts, and other difficult machining process. There are unique advantages of ECM over other micro-machining techniques, such as no heat-affected zone, no tool wear, independent of workpiece hardness, and high surface finish, which is a contactless process for metallic material dissolution [12,13].

Jet electrochemical machining (JEM) sprays high-speed electrolyte from the metal nozzle to the surface of the workpiece, which helps to generate deep micro-grooves, since the velocity of the electrolyte and the mass transfer effect of the machining process are improved. Natsu et al. [14] developed an algorithm to control the scanning speed of the nozzle by superimposition and generated micro-groove arrays on the plane and the cylindrical surface. Hackert-Oschätzchen et al. [15] applied the JEM process to fabricate microchannels with a width of 200 μm and a depth of 60 μm by controlling the trajectory of nozzle with inner diameter of 100 μm. Mitchell-Smith et al. [16,17] explored the difference in groove profile and resultant surface finish by changing the inclination angle of the nozzle. However, JEM always suffers from stray corrosion at the edges of micro-grooves and has limitations in micromanufacturing due to hydraulic jump phenomenon and difficulty in fabricating suitable nozzles with small inner diameter [18]. Through-mask electrochemical micromachining (TMEMM) is also an effective approach to fabricate surface microstructures. In the standard TMEMM, the surface of each workpiece is coated with a patterned mask prepared by the lithography process. Based on the electrochemical reaction, the exposed regions of the workpiece surface are dissolved [19]. Zhou et al. [20] employed the TMEMM process to generate micro-grooves ranging 148.5 to 258.6 μm in width and 17.6 to 58.0 μm in depth on the surface of phosphor bronze alloy. They used a mask with a 110 μm wide micro-slit and studied the effect of different machining parameters on the dimensions and morphology of micro-grooves. Overcut in TMEMM is generated inevitably due to the isotropic nature of metal dissolution. It should be noted that the photoresist on the workpiece surface needs to be removed after TMEMM, and therefore it is not reusable. Zhu et al. [21] developed a modified TMEMM process, which replaced the photoresist layer with a re-usable mask and reduced the processing cost. Qu et al. [22] proposed a modified microscale pattern transfer by employing a re-usable cathodic tool carrying the dry-film mask. Wang et al. [23] explored the applicability of electrochemical pattern transfer machining (ECPTM) to fabricate micro-grooves and performed a simulation analysis. It is worth pointing out that ECPTM also suffers from severe stray corrosion on both sides of the micro-groove due to the unprotected surface of the workpiece. Zhang et al. [24] developed a sandwich-like electrochemical micromachining (SLEMM) technology for reducing overcut and improving the dimensional uniformity of micro-dimples by enabling uniform distribution of the electric field in the machining region. However, the generation of bubbles and accumulation of insoluble products in the machining region suppress the dissolution of the workpiece.

In order to evacuate bubbles and some insoluble electrolytic products as well as optimize the cathode structure of conventional SLEMM, Zhang et al. [25] proposed a process scheme of the open reaction unit by using a porous cathode. Ming et al. [26] analyzed and verified the electric field distribution characteristics of the above method and fabricated micro-dimples arrays of dimensional uniformity on the cylindrical workpiece surface. Stray corrosion is an unavoidable phenomenon in TMEMM. The auxiliary anode was proposed to reduce the marginal effect of the electric field on the workpiece, thereby improving the machining accuracy of the microstructure [27]. Chen et al. [28] developed a method of JEM with a conductive mask to generate micro-dimples, which evidently reduced the overcut and enhanced the machining localization.

The literature has shown the impact of processing technology on the fabrication of microstructures. For the electrochemical machining of micro-grooves, the key to improve the machining quality is to enhance the mass transfer effect in the machining region and reduce overcut to heighten the localization. In previous research [29], we have proposed a method of electrochemical machining of micro-grooves with masked porous cathode, the influence of different flow modes on the machining process was investigated, and a jet flow mode was optimized at last, which could improve the machining efficiency and dimensional uniformity. However, the undercut of micro groove is unavoidable due to the isotropy of material when the traditional insulated mask was used, which reduced the machining localization. In this paper, a conductive masked porous cathode with jet electrolyte supply is proposed (Figure 1) for the fabrication of micro-grooves. The conductive mask is introduced to be directly and firmly attached to the workpiece, replacing the original insulated mask, so that the conductive mask had the same potential as the workpiece. This will weaken the electric field intensity on both sides of the micro-grooves, thereby achieving the purpose of reducing overcut. Moreover, a metallic nozzle is used to increase the flow rate of the electrolyte in the machining region and enhance mass transfer. In this paper, the distribution of the electric field in the machining region is analyzed by establishing a simulation model with different masks, and the evolution processes of the micro-groove profile is predicted as well. The feasibility of the method is reported. Combining with the experimental results, the effect of electrolyte pressure and pulse duty cycle on the fabrication of micro-grooves are analyzed.

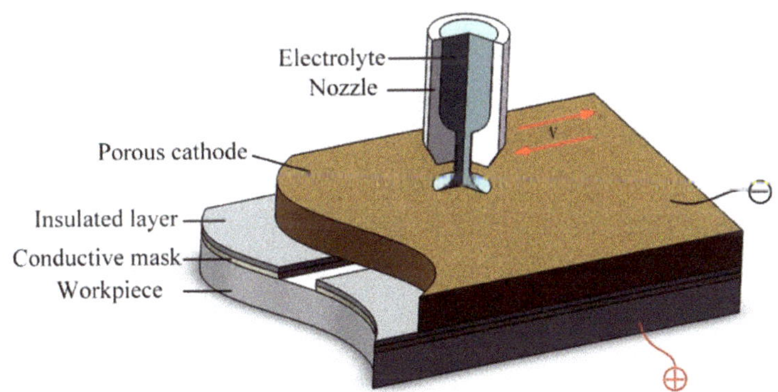

Figure 1. The schematic diagram of jet machining with a conductive masked porous cathode.

2. Description of the Method and Numerical Simulation

2.1. Description of the Method

Due to relatively larger machining gap in the conventional TMEMM, the mass transfer is promoted. On the other hand, by minimizing the inter-electrode gap, the electric field distribution between the electrodes is uniformly distributed, and the dimensional uniformity is improved. Figure 1 shows a schematic diagram of the jet machining of micro-grooves by using a conductive masked porous cathode. One side of the insulated layer is attached with the porous metal, while the other side is bonded to the conductive mask as a whole, which is covered on the workpiece, and thus the machining region is restrained in a closed unit. The porous metal is used as a cathode and mass transport medium. The columnar pressurized electrolyte is ejected toward the porous cathode from the nozzle, which can provide a high electrolyte velocity to infiltrate into the whole machining region through the porous cathode. A micro-groove can be generated when sufficient voltage is applied between the porous cathode and workpiece. The processing condition in the whole machining region can be improved by controlling the reciprocating motion of the nozzle, which is helpful for the removal of the electrolytic product as well as for the renewal of the electrolytes. After processing, the porous cathode and mask can be re-used, and it could reduce the processing cost as well as improve the machining efficiency.

2.2. Numerical Simulation

In this study, both the porous cathode and the conductive mask were introduced. For comparative analysis, the electric field distribution in the machining region was investigated with both insulated and conductive masks by establishing a mathematical simulation model, and the evolved shape of the micro-groove was predicted.

2.2.1. Model Building

Since only small potential gradients are expected in the metallic electrodes due to the high conductivities, the electrode domains are not included in the model. The insulated layer is electrochemically inert, and hence it is not included either. The only modeled domain is the electrolyte. By simplifying the process model, the 2D electric field simulation model of the machining region is established in the inter-electrode gap, as shown in Figure 2. H_1 and H_2 denote the thickness of insulated layer and mask, respectively. W is the width of the micro-slot of the mask. Considering that the anode voltage during actual processing is distributed over the entire workpiece, a small virtual area with a length of 400 µm and height of 5 µm is added at anode end, which also makes boundary 6 move properly without mesh distortion.

According to electric field theory, the electric potential φ is governed by Laplace's equation [30]:

$$\nabla^2 \varphi = 0. \tag{1}$$

According to Ohm's law, the current density, $\vec{i}$, can be evaluated as follows:

$$\vec{i} = \sigma \vec{E} \tag{2}$$

where σ is the electrolyte conductivity and $\vec{E}$ is the electric field intensity on the workpiece.

According to Faraday's Law, the rate of material dissolution, $\vec{v}$, can be expressed as follows:

$$\vec{v} = \eta \omega \vec{i} = \eta \omega \sigma \vec{E} \tag{3}$$

where η represents the current efficiency and ω represents the volumetric electrochemical equivalent of the material.

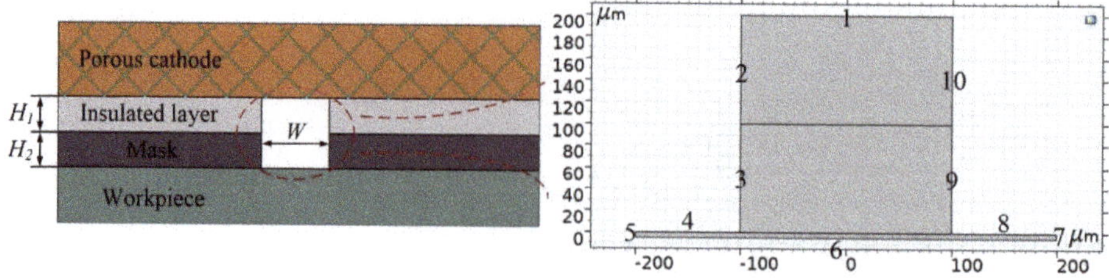

Figure 2. The 2D model of electrochemical machining (ECM) with a masked porous cathode.

In this study, the current efficiency can be expressed as follows [31]:

$$\eta(i) = \frac{0.85}{\left(1 + e^{(10-i)/6}\right)} - 0.1. \tag{4}$$

To simplify the model of the electric field established in the machining process, the following assumptions are made:

1. The reaction surface of porous cathode is considered as a solid flat surface, ignoring the internal porous structure.
2. The potential loss due to the electrochemical reaction is ignored, and Ohm's law and charge balance are used to calculate the current in the electrolyte and electrode.
3. The conductivity of the electrolyte, σ, is constant.
4. The concentration gradient in the bulk electrolyte is negligible.

In the model, the primary current distribution is used to calculate the electric field intensity. The domain of computational interest in the model is the zone of electrolyte. Boundary 1, whose external potential is set to 0 V (ground), is the cathode reaction surface (porous cathode surface). Boundary 6 is the anode reaction surface (workpiece surface), which is dissolved during the electrochemical reaction process, and its external potential is set to U. With a conductive mask, Boundaries 3–4 and 8–9 are set to the electric potential of U. Correspondingly, they are set to electric insulation with an insulated mask, and the other undefined boundaries are set as electric insulation.

By coupling the primary current distribution and deformed geometry module, the process of material dissolution was simulated. Combining Equations (3) and (4), deformation of Boundary 6 can be obtained. Then, the evolved shape of the micro-groove profile is described. In this model, the boundaries are set as follows:

$$\begin{cases} \frac{d_x}{d_t}|\Gamma_{2,3,5,7,9,10} = 0 \\ \frac{d_y}{d_t}|\Gamma_{1,4,8} = 0 \\ \frac{\partial(x,y)}{\partial_n\partial_t}|\Gamma_6 = v_n \end{cases} \tag{5}$$

where v_n represents the normal anodic dissolution rate on the workpiece in Boundary 6.

The parameters used in the simulation are listed in Table 1. The model is solved on a commercial finite element platform COMSOL® Multiphysics software version 5.4.

Table 1. The parameters set for the simulation.

Parameters	Value
Thickness of the insulated layer, H_1	100 μm
Thickness of the mask, H_2	100 μm
Width of the micro-slot, W	200 μm
Electrolyte conductivity, σ	12 S/m
Volumetric electrochemical equivalent, ω	0.035 mm³/(A·s)
Electric potential, U	35 V

2.2.2. Simulation Results

The initial electric field distribution on the exposed workpiece surface simulated using conductive and insulated masks is shown in Figure 3. It can be seen that the electric field intensity using an insulated mask is greater than that obtained with a conductive mask at each position. The electric field intensity using the conductive mask is gradually increased from the edge to the center in the machining region. Correspondingly, with the insulated mask, the electric field intensity is quickly transitioned from the minimum value at the edge to the maximum constant value in the middle region. By observing the magnitude of the electric field intensity value, it can be inferred that the dissolution rate of the material is proportional to the electric field intensity according to Equations (2)–(4). In process terms, this means that more material is dissolved at the region toward edges with the insulated mask, while this is not the case when the conductive mask is used.

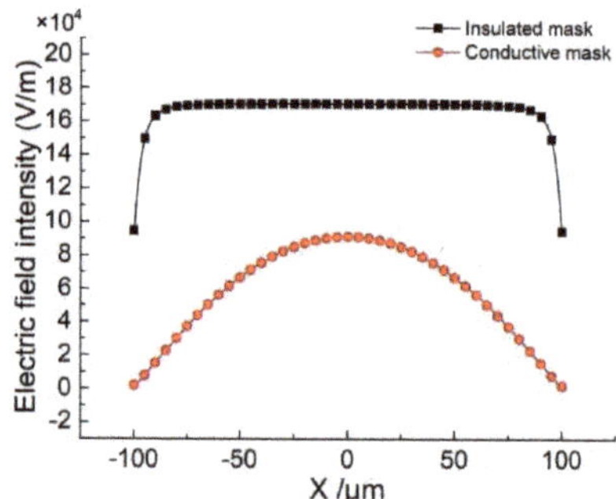

Figure 3. Electric field intensity on the workpiece surface using different masks.

Figure 4 shows the simulation results for grooves generated with the same depth of 45 μm using conductive and insulated masks, respectively. It can be seen that while using an insulated mask, the groove is much wider than those obtained with a conductive mask. This is a strong indication that the machining localization can be enhanced with the conductive mask.

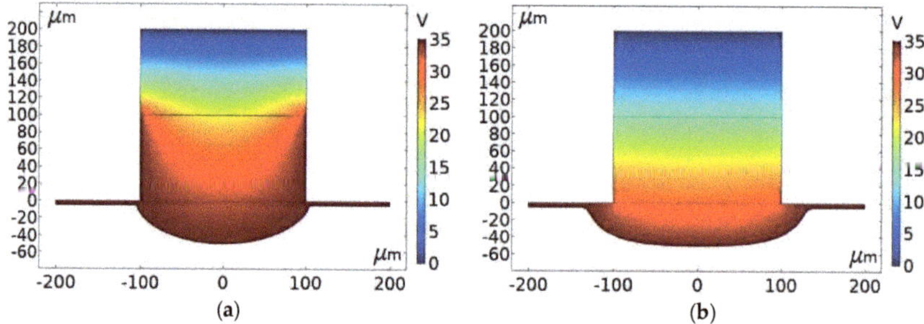

Figure 4. Simulation results using different masks. (**a**) Using the conductive mask; (**b**) Using the insulated mask.

In order to further analyze the differences in the micro-grooves fabricated using different masks, simulations were conducted at different depths to compare the change in width. Figure 5 shows cross-section profiles of micro-grooves generated at different depths (15 μm, 25 μm, 35 μm, and 45 μm). The profiles also reflect the evolution process of micro-grooves with a conductive mask and insulated mask. As the depth of the micro-grooves fabricated with an insulated mask is increased, the width also increases. Correspondingly, with a conductive mask, there was only a small increase in the width, even at higher depths. Moreover, at the same depth, the width of micro-grooves generated with a conductive mask was smaller than that of using an insulated mask. The phenomenon can be explained as follows.

The electric field is confined to a closed machining region with the proposed method. For the use of the insulated mask, the electric field is formed directly by a porous cathode and workpiece anode. It can be seen from Figure 3 that the electric field intensity is larger compared with the conductive mask. According to Equation (3), although the electric field intensity at the center of the machining region is much greater than the edges, the material of the width direction is also dissolved to a certain extent with the increase of depth. Correspondingly, for the use of the conductive mask, the conductive mask has the same potential as the workpiece anode, so that the electric field also is formed by a conductive mask and porous cathode, which weakens the electric field intensity on the workpiece edge surface of the machining region. Furthermore, the width growth range is even smaller compared with the depth increasing. Through comparative analysis, the use of a conductive mask is more helpful to reduce the overcutting and improve the machining localization.

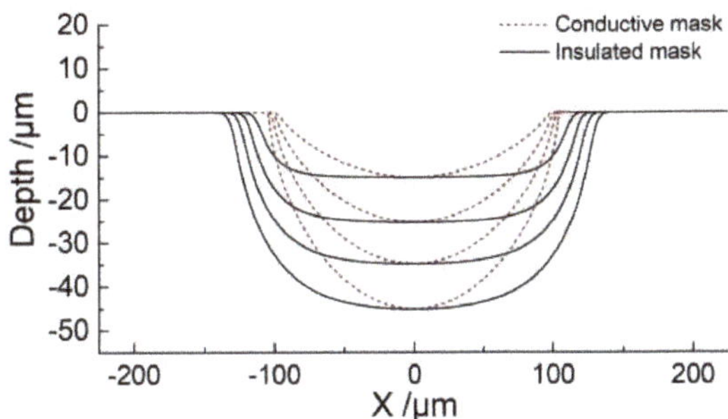

Figure 5. Cross-section profiles of micro-grooves with different masks.

3. Experimental

The developed experimental system is schematically shown in Figure 6. It consists of a servo-control feed unit, platform control unit, electrolyte circulation system, and power supply. The conductive porous cathode and workpiece were fixed on the X/Y stage by a special fixture. The nozzle was clamped on the Z-axis by the holder. The motion route and speed of the nozzle were controlled along the position of the through-mask above the porous cathode. The electrolyte circulating system could realize the reuse of an electrolyte and ensure the refreshing of gap conditions through the recirculation of an electrolyte. In the end, the micro-groove could be fabricated by applying a sufficient voltage.

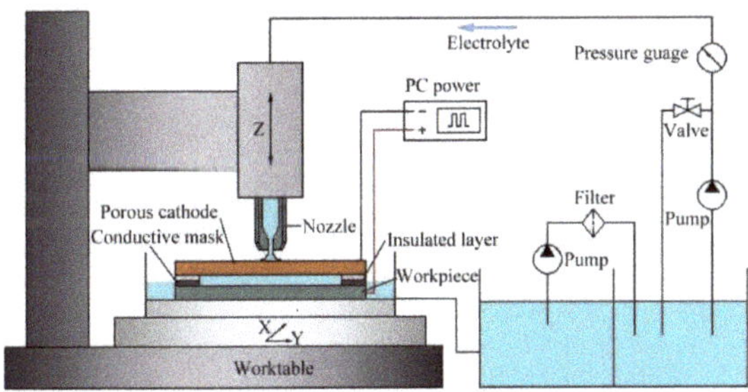

Figure 6. Schematic of an experimental system.

In this experiment, a porous copper plate was introduced as a porous cathode (Figure 7a), and a platinum sheet with a thickness of 100 μm was employed as the conductive mask, which showed a high chemical inertness and could not be dissolved during machining. The micro-slot with a length of 20 mm and a width of 200 μm in the mask was prepared by a micro-electrical discharge process (shown in Figure 7b). The flexible insulated layer was made of epoxy resin with a thickness of 100 μm. During the electrochemical machining, the nozzle was reciprocated once, and a nozzle with an inner diameter of 2 mm was utilized. The nozzle is moved 1 mm away from the porous cathode. The main experiment parameters are listed in Table 2. In our previous research, we have found that increasing the reciprocating motion number (increasing the moving speed of nozzle) could reduce the machining roughness in high machining voltage (high current density), but it has little influence on the machining roughness in low voltage (low current density), because there was less electrolytic product [29]. According to the simulation result in Figure 3, it is found that with a conductive mask, the electric field intensity on the workpiece is quite lower than that with an insulated mask, which means a low current density on workpiece during machining. According to the previous research result, the influence of the reciprocating motion number on the roughness would not be obvious due to the low current density, and our pre-experiments have also proved it, thus one reciprocating motion of a nozzle was used in this study.

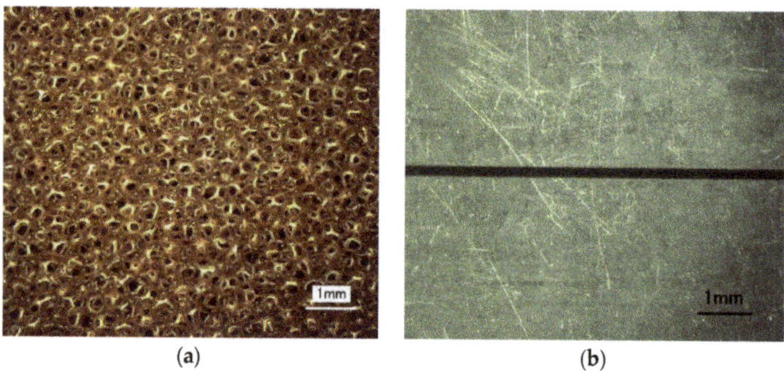

Figure 7. The image of conductive masked porous cathode. (**a**) Porous cathode; (**b**) Conductive mask.

Table 2. Experimental machining parameters.

Parameters	Value
Electrolyte concentration	12% (wt %), $NaNO_3$
Electrolyte temperature, T	25 °C
Inner diameter of nozzle, d	2 mm
Electrolyte pressure, P_{in}	0.2, 0.4, 0.6, 0.8 MPa
Thickness of the porous cathode, T_1	3 mm
Porosity of the porous cathode, ε_p	0.95
Thickness of the insulated layer, T_2	100 μm
Thickness of the conductive mask, T_3	100 μm
Length of the micro-slit, L	20 mm
Width of the micro-slit, W	200 μm
Reciprocating motion number of the nozzle, N	1
Applied voltage with the insulated mask, U_1	10 V
Applied voltage with the conductive mask, U_2	35 V
Pulse duty cycle, ε	20%, 40%, 60%, 80%
Pulse frequency, f	1 kHz
Machining time with the insulated mask (t_{on}), t_1	10 s
Machining time with the conductive mask (t_{on}), t_2	20 s
Workpiece material	Stainless steel 304
Metallic nozzle material	Stainless steel 304

A hall current sensor was used to detect the current signal, and a data acquisition card (NI 9222, Austin, TX, USA) was introduced to record current with a sampling frequency of 100 kHz. In order to analyze the repeatability of fabrication of a micro-groove with diffident parameters, each set of experimental parameters was repeated twice. Seven points were measured for each micro-groove, the average value of the two micro-grooves was obtained, and the standard deviation (SD) was used to evaluate the dimensional uniformity of a micro-groove. The profiles and bottom roughness of micro-grooves were measured using a confocal laser-scanning microscope (CLSM, Olympus LEXT OLS4000, Tokyo, Japan).

4. Results and Discussion

4.1. Comparison of Micro-Grooves Generated with Conductive and Insulated Mask

For the purpose of comparing the difference in the fabrication of micro-grooves with different masks, the experiments were conducted with pulse power supply setting a pulse duty cycle of 20% and pulse frequency of 1 kHz. Among them, the applied voltage of 10 V and machining time of 10 s with the insulated mask were adopted, and the applied voltage of 35 V and machining time of 20 s with the conductive mask were employed. It should be pointed out that the use of a conductive mask weakens the electric field intensity in the machining region, requiring a larger voltage during processing. The insulated mask and conductive mask with thicknesses of 100 μm were designed with same structure. The width and length of the micro-slot in the masks were 200 μm and 20 mm, respectively.

Figure 8 shows profiles of fabricated micro-grooves with same depth of 45 μm using insulated and conductive masks. It can be seen that the width of the micro-groove was about 319.4 μm by using an insulated mask (Figure 8a). However, when a conductive mask was introduced, the width of the micro-groove was about 227.3 μm (Figure 8b). This is an indication that overcut is significantly improved with a conductive mask as compared to using the insulated mask. The etch factor, which is defined as depth/(width of groove—width of mask), increases from 0.75 to 3.3, and it reflects that the machining localization of the micro-groove is enhanced. Although the experimental profiles show good agreement with the simulation results, the undercut of its width is greater than the simulation result, which reduces the machining accuracy. The main reason for this observation can be that the mask cannot be adhered firmly to the workpiece, which causes the electrolyte to penetrate into the gap between the mask and the workpiece, resulting in overcut along the width.

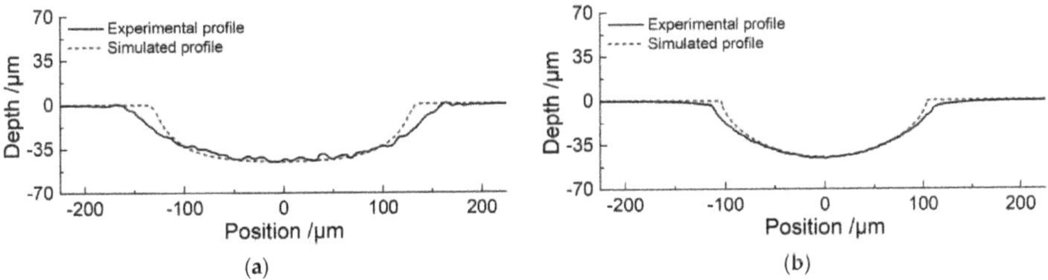

Figure 8. Profiles of micro-grooves in same depth fabricated with different masks. (a) Using the insulated mask (P_{in} = 0.8 MPa, U = 10 V, ε = 20%, f = 1 kHz, t_{on} = 10 s); (b) Using the conductive mask (P_{in} = 0.8 MPa, U = 35 V, ε = 20%, f = 1 kHz, t_{on} = 20 s).

It is clear from the above discussions that the undercut using the conductive mask is smaller than that using the insulating mask, showing that the conductive mask can reduce the electric field intensity on both sides of the machining region. On the other hand, the use of a conductive mask requires a higher voltage and a longer time, which reflects that the energy consumption was increased. The following experiments were performed to study the effect of electrochemical machining of micro-grooves using a conductive masked porous cathode and jetted electrolyte.

4.2. Micro-Grooves Generated with Different Electrolyte Pressures

In order to investigate the profile of micro-grooves generated at different electrolyte pressure, the experiments were prepared with a PC voltage of 35 V, pulse duty cycle of 20%, pulse frequency of 1 kHz, and machining time of 20 s. Figure 9 shows the cross-section profiles, the 3D profiles, and the bottom profiles of the micro-grooves. It can be seen that the dimensional uniformity of the micro-groove of jet machining using the conductive masked porous cathode is good. As the electrolyte pressure is increased, profiles of micro-grooves become smoother.

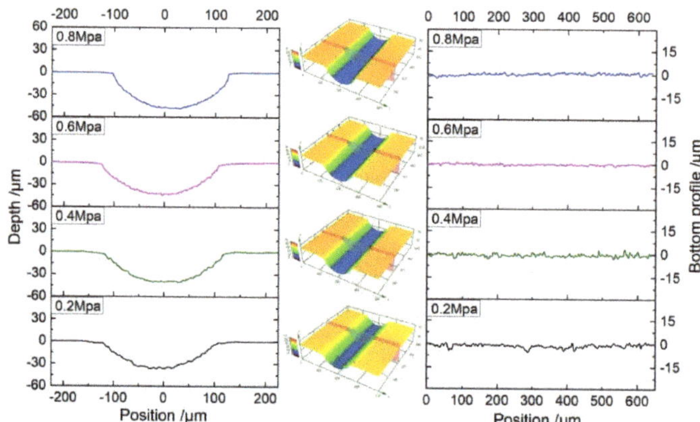

Figure 9. The profiles of micro-grooves generated with different electrolyte pressures ($U = 35$ V, $\varepsilon = 20\%$, $f = 1$ kHz, $t_{on} = 20$ s).

Figures 10 and 11 show the dimension and roughness of micro-grooves fabricated at different electrolyte pressures, respectively. When the electrolyte pressure is increased from 0.2 MPa to 0.8 MPa, the width increases from 237.3 ± 4.8 μm to 232.6 ± 5.7 μm (mean ± SD), the depth increases from 39.2 ± 4 μm to 46.9 ± 3.9 μm, and the roughness (Ra) decreases from 1.13 μm to 0.39 μm. It shows that with an increase in electrolyte pressure, the electrolyte pressure helps to make the conductive masked porous cathode adhere better to the workpiece. This reduces the penetration of electrolyte between the mask, and the workpiece and closed electric field is maintained. Consequently, grooves with smaller width and higher depth are obtained, and the standard deviation is maintained at a smaller value. In the case of low electrolyte pressure of 0.2 MPa and 0.4 MPa, the mass transfer of electrolyte became relatively weakened due to the infiltration of insoluble reaction by-products as well as gas bubbles into the porous cathode. This suppressed further electrochemical machining resulting in a poor flat uniformity of the bottom profile along the length direction. Compared with electrolyte pressure of 0.2 MPa and 0.4 MPa, a high electrolyte velocity could be provided around the machining region in the applied electrolyte pressure of 0.6 MPa and 0.8 MPa, which was helpful for refreshing the electrolyte and the evacuation of the insoluble products, bubbles, and heat. The electrochemical machining environment is optimized by reciprocating the nozzle, which could increase the stability of the machining process. Therefore, the prepared micro-groove shows small roughness and high machining quality. The electrolyte pressure was selected to 0.8 MPa in the following experiments.

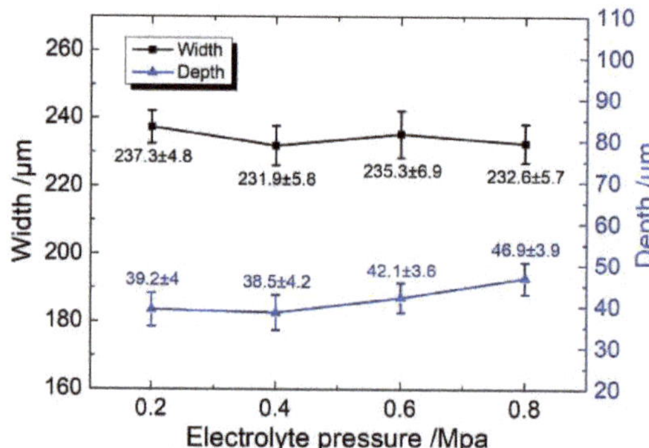

Figure 10. The dimension of micro-grooves generated with different electrolyte pressures ($U = 35$ V, $\varepsilon = 20\%$, $f = 1$ kHz, $t_{on} = 20$ s).

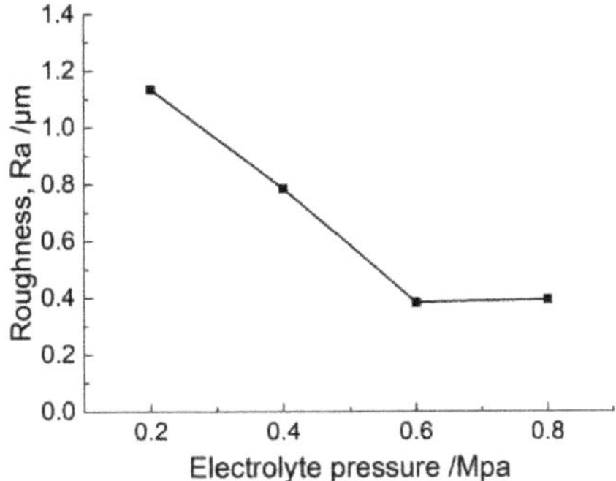

Figure 11. The roughness of micro-grooves generated with different electrolyte pressures ($U = 35$ V, $\varepsilon = 20\%$, $f = 1$ kHz, $t_{on} = 20$ s).

4.3. Micro-Grooves Generated with Different Pulse Duty Cycles

Besides the electrolyte pressure, the pulse duty cycle also has a significant effect on the fabrication of micro-grooves. The pulse current is helpful to improve the machining accuracy of ECM and to refresh the electrolyte during the pulse-off time. In the above study, the pulse duty cycle of 20% was utilized. In order to investigate the influence of the pulse duty cycle on the machining process, experiments with different pulse duty cycles were conducted at a voltage of 35 V.

Figure 12 shows the dimensions of the micro-grooves fabricated with different pulse duty cycles at a frequency of 1 kHz and for the same machining time of 20 s. It can be seen that there is an obvious change in the width of the micro-grooves with an increasing pulse duty cycle. The width increases from 232.6 ± 5.7 μm to 274.1 ± 12.5 μm when the pulse duty cycle increases from 20% to 80%. On the other hand, the depth of the micro-groove showed a slight change from 46.9 ± 3.9 μm to 53.9 ± 4.1 μm.

As we know, the electrochemical reactions cause heat generated in the bulk of the electrolyte (Joule heating) during machining. Especially with a conductive mask, there is a larger reaction area than that in insulated mask, and there would be a large amount of heat generated. This heat increases the electrolyte temperature, resulting in a further increase of the electrolyte conductivity. In the pulse current, the pulse off time is used to flush the heat and by-products out of the machining gap as well as renew the electrolyte. Consequently, the influence of heat on the machining process is reduced. When the pulse duty cycle is increased, the pulse-off time is reduced. As a result, the heat transport process in the machining gap is weakened, which increases the electrolyte conductivity. According to Equation (3), the material removal rate is increased with a high pulse duty cycle. Figure 12 shows that with the increasing pulse duty cycle, both the width and depth are enlarged. On the other hand, the decrease of pulse-off time weakens the removal of electrolytic products in the gap. As a result, the accumulation of electrolytic products at the bottom of the micro-groove is increased, and it hinders the material dissolution at the bottom of the micro-groove. That is why the increase of the material removal rate is more in the lateral direction (width) as compared to the axial direction (depth). Meanwhile, the accumulation of electrolytic products led to a non-uniform dissolution of material in both width and depth, and it not only increased the error bar, but also reduced the surface quality. Figures 13 and 14 show the profiles and roughness of micro-grooves fabricated with different pulse duty cycles, respectively. It can be observed that as the duty cycle is reduced from 80% to 20%, the surface roughness (Ra) reduces to 0.393 μm. The localization and surface quality of micro-grooves is improved with a low pulse duty cycle. Therefore, the pulse duty cycle of 20% is found to be more suitable for the preparation of micro-grooves by this process scheme.

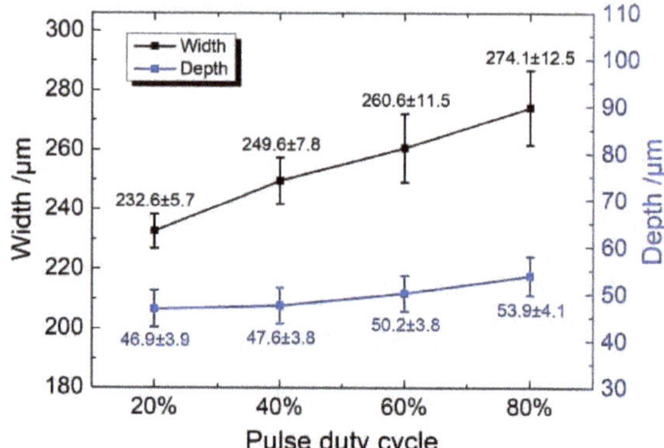

Figure 12. The dimension of micro-grooves generated with different pulse duty cycles (P_{in} = 0.8 MPa, U = 35 V, f = 1 kHz, t_{on} = 20 s).

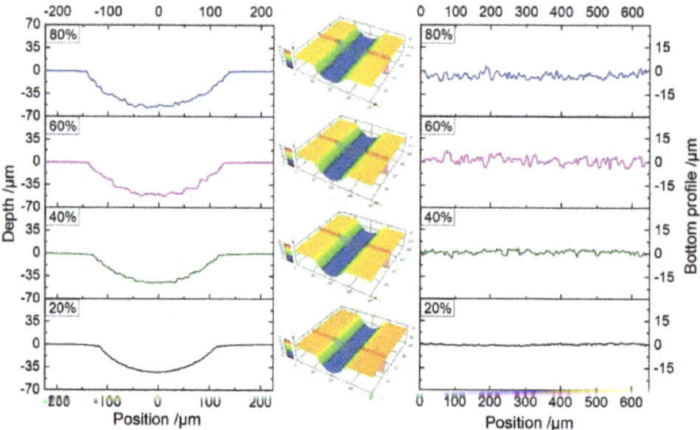

Figure 13. The profile of micro-grooves generated with different pulse duty cycles (P_{in} = 0.8 MPa, U = 35 V, f = 1 kHz, t_{on} = 20 s).

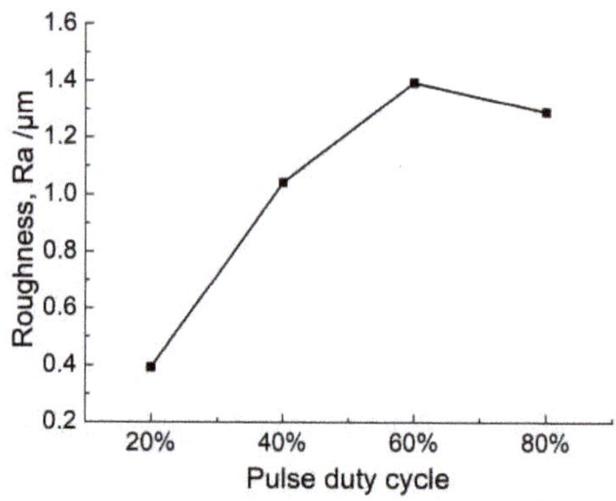

Figure 14. The roughness of micro-grooves with different pulse duty cycles (P_{in} = 0.8 MPa, U = 35 V, f = 1 kHz, t_{on} = 20 s).

5. Conclusions

This paper proposed a method of electrochemical machining of micro-grooves using a conductive masked porous cathode and jet electrolyte supply. From the simulation and experimental results, the concluding remarks can be summarized as follows.

1. The simulation results indicated that the use of a conductive mask reduced the electric field intensity on both sides of the micro-groove and achieved the purpose of reduction in overcut.
2. On comparing the results with an insulated mask and a conductive mask to fabricate micro-grooves with the same depth of 45 μm, it was observed that the etch factor was increased from 0.75 (insulated mask) to 3.3 (conductive mask), which showed that the machining localization of micro-grooves was enhanced.
3. In this process scheme, a high electrolyte pressure was favorable for the renewal of the electrolyte and enhanced mass transfer during processing, which improved the machining quality and dimensional uniformity of the micro-grooves.
4. The pulse duty cycle has an important effect on the machining localization. A low pulse duty cycle of 20% could obtain micro-grooves with better machining localization and surface quality.

Author Contributions: Conceptualization, X.C.; methodology, X.C. and G.F.; software, X.C., J.L., and G.F.; validation, X.C., G.F., and K.K.S.; formal analysis, X.C. and K.K.S.; investigation, X.C., J.L., and G.F.; resources, X.C., J.L., and Z.G.; data curation, G.F.; writing—original draft preparation, G.F.; writing—review and editing, X.C., J.L., and K.K.S.; supervision, X.C. and J.L.; project administration, X.C. and Z.G.; funding acquisition, X.C., J.L. and Z.G. All authors have read and agreed to the published version of the manuscript.

Funding: This research was funded by The Joint Funds of the National Natural Science Foundation of China and Guangdong Province (Grant No. U1601201), National Natural Science Foundation of China (Grant No. 51705089, 51675105), and Pearl River S and T Nova Program of Guangzhou (Grant No. 201906010099).

Conflicts of Interest: The authors declare no conflict of interest.

References

1. Uhlmann, E.; Mullany, B.; Biermann, D.; Rajurkar, K.P.; Hausotte, T.; Brinksmeier, E. Process chains for high-precision components with micro-scale features. *CIRP Ann. Manuf. Technol.* **2016**, *65*, 549–572. [CrossRef]
2. Bruzzone, A.A.G.; Costa, H.L.; Lonardo, P.M.; Lucca, D.A. Advances in engineered surfaces for functional performance. *CIRP Ann. Manuf. Technol.* **2008**, *57*, 750–769. [CrossRef]
3. Fontana, É.; Mancusi, E.; da Silva, A.; Mariani, V.C.; Ulson de Souza, A.A.; Ulson de Souza, S.M.A.G. Study of the effects of flow channel with non-uniform cross-sectional area on PEMFC species and heat transfer. *Int. J. Heat Mass Transf.* **2011**, *54*, 4462–4472. [CrossRef]
4. Bozorgnezhad, A.; Shams, M.; Kanani, H.; Hasheminasab, M.; Ahmadi, G. The experimental study of water management in the cathode channel of single-serpentine transparent proton exchange membrane fuel cell by direct visualization. *Int. J. Hydrog. Energy* **2015**, *40*, 2808–2832. [CrossRef]
5. Mishra, S.K.; Ghosh, S.; Aravindan, S. Characterization and machining performance of laser-textured chevron shaped tools coated with AlTiN and AlCrN coatings. *Surf. Coat. Technol.* **2018**, *334*, 344–356. [CrossRef]
6. da Silva, W.M.; Suarez, M.P.; Machado, A.R.; Costa, H.L. Effect of laser surface modification on the micro-abrasive wear resistance of coated cemented carbide tools. *Wear* **2013**, *302*, 1230–1240. [CrossRef]
7. Xiang, G.; Han, Y.; Wang, J.; Xiao, K.; Li, J. A transient hydrodynamic lubrication comparative analysis for misaligned micro-grooved bearing considering axial reciprocating movement of shaft. *Tribol. Int.* **2019**, *132*, 11–23. [CrossRef]
8. Chang, D.Y.; Shen, P.C.; Hung, J.C.; Lee, S.J.; Tsui, H.P. Process Simulation–Assisted Fabricating Micro-Herringbone Grooves for a Hydrodynamic Bearing in Electrochemical Micromachining. *Mater. Manuf. Process.* **2011**, *26*, 1451–1458. [CrossRef]
9. Shamoto, E.; Saito, A. A novel deep groove machining method utilizing variablepitch end mill with feed-directional thin support. *Precis. Eng.* **2016**, *43*, 277–284. [CrossRef]

10. Schreck, S.; Zum, G.K.H. Laser-assisted structuring of ceramic and steel surfaces for improving tribological properties. *Appl. Surf. Sci.* **2005**, *247*, 616–622. [CrossRef]
11. Hung, J.C.; Chang, D.H.; Chuang, Y. The fabrication of high-aspect-ratio microflow channels on metallic bipolar plates using die-sinking micro-electrical discharge machining. *J. Power Sources* **2012**, *198*, 158–163. [CrossRef]
12. Saxena, K.K.; Qian, J.; Reynaerts, D. A review on process capabilities of electrochemical micromachining and its hybrid variants. *Int. J. Mach. Tools Manuf.* **2018**, *127*, 28–56. [CrossRef]
13. Fang, X.; Wang, X.; Wang, W.; Qu, N.; Li, H. Electrochemical drilling of multiple small holes with optimized electrolyte dividing manifolds. *J. Mater. Process. Technol.* **2017**, *247*, 40–47. [CrossRef]
14. Natsu, W.; Ikeda, T.; Kunieda, M. Generating complicated surface with electrolyte jet machining. *Precis. Eng.* **2007**, *31*, 33–39. [CrossRef]
15. Hackert-Oschätzchen, M.; Meichsner, G.; Zinecker, M.; Martin, A.; Schubert, A. Micro machining with continuous electrolytic free jet. *Precis. Eng.* **2012**, *36*, 612–619. [CrossRef]
16. Mitchell-Smith, J.; Speidel, A.; Clare, A.T. Advancing electrochemical jet methods through manipulation of the angle of address. *J. Mater. Process. Technol.* **2018**, *255*, 364–372. [CrossRef]
17. Mitchell-Smith, J.; Speidel, A.; Gaskell, J.; Clare, A.T. Energy distribution modulation by mechanical design for electrochemical jet processing techniques. *Int. J. Mach. Tools Manuf.* **2017**, *122*, 32–46. [CrossRef]
18. Ikeda, T.; Natsu, W.; Kunieda, M. Electrolyte jet machining using multiple nozzles. *Int. J. Electr. Mach.* **2006**, *11*, 25–32. [CrossRef]
19. Hao, X.; Wang, L.; Wang, D.; Guo, F.; Tang, Y.; Ding, Y.; Lu, B. Surface micro-texturing of metallic cylindrical surface with proximity rolling-exposure lithography and electrochemical micromachining. *Appl. Surf. Sci.* **2011**, *257*, 8906–8911. [CrossRef]
20. Zhou, X.; Qu, N.; Hou, Z.; Zhao, G. Electrochemical micromachining of microgroove arrays on phosphor bronze surface for improving the tribological performance. *Chin. J. Aeronaut.* **2018**, *31*, 1609–1618. [CrossRef]
21. Zhu, D.; Qu, N.S.; Li, H.S.; Zeng, Y.B.; Li, D.L.; Qian, S.Q. Electrochemical micromachining of microstructures of micro hole and dimple array. *Cirp Ann. Manuf. Technol.* **2009**, *58*, 177–180. [CrossRef]
22. Qu, N.S.; Zhang, X.F.; Chen, X.L.; Li, H.S.; Zhu, D. Modified microscale pattern transfer without photolithography of substrates. *J. Mater. Process. Technol.* **2015**, *218*, 71–79. [CrossRef]
23. Wang, Z.; Qian, S.; Cao, H.; Zhang, H.; Bao, K. Simulation and Analysis on Machining Channels in the Electrochemical Pattern Transfer Method. *Int. J. Electrochem. Sci.* **2016**, *11*, 6126–6137. [CrossRef]
24. Zhang, X.; Qu, N.; Chen, X. Sandwich-like electrochemical micromachining of micro-dimples. *Surf. Coat. Technol.* **2016**, *302*, 438–447. [CrossRef]
25. Zhang, X.; Qu, N.; Fang, X. Sandwich-like electrochemical micromachining of micro-dimples using a porous metal cathode. *Surf. Coat. Technol.* **2017**, *311*, 357–364. [CrossRef]
26. Ming, P.; Zhao, C.; Zhang, X.; Li, X.; Qin, G.; Yan, L. Investigation of foamed cathode through-mask electrochemical micromachining developed for uniform texturing on metallic cylindrical surface. *Int. J. Adv. Manuf. Technol.* **2018**, *96*, 3043–3056. [CrossRef]
27. Chen, X.; Qu, N.; Li, H. Improvement of dimensional uniformity on micro-dimple arrays generated by electrochemical micro-machining with an auxiliary electrode. *Int. J. Adv. Manuf. Technol.* **2015**, *80*, 1577–1585. [CrossRef]
28. Chen, X.L.; Dong, B.Y.; Zhang, C.Y.; Wu, M.; Guo, Z.N. Jet electrochemical machining of micro dimples with conductive mask. *J. Mater. Process. Technol.* **2018**, *257*, 101–111. [CrossRef]
29. Chen, X.L.; Fan, G.C.; Lin, C.H.; Dong, B.Y.; Guo, Z.N.; Fang, X.L.; Qu, N.S. Investigation on the electrochemical machining of micro groove using masked porous cathode. *J. Mater. Process. Technol.* **2020**, *276*, 116406. [CrossRef]
30. Wang, D.; Zhu, Z.; Bao, J.; Zhu, D. Reduction of stray corrosion by using iron coating in NaNO$_3$ solution during electrochemical machining. *Int. J. Adv. Manuf. Technol.* **2015**, *76*, 1365–1370. [CrossRef]
31. Fang, X.; Qu, N.; Zhang, Y.; Xu, Z.; Zhu, D. Effects of pulsating electrolyte flow in electrochemical machining. *J. Mater. Process. Technol.* **2014**, *214*, 36–43. [CrossRef]

 *micromachines*

Article

The Optimal Processing Parameters of Radial Ultrasonic Rolling Electrochemical Micromachining—RSM Approach

Kailei He, Xia Chen and Minghuan Wang * 

Key Laboratory of Special Purpose Equipment and Advanced Processing Technology,
Ministry of Education & Zhejiang Province, Zhejiang University of Technology, Hangzhou 310014, China;
hekailei@zjut.edu.cn (K.H.); cx33150@163.com (X.C.)
* Correspondence: wangmh@zjut.edu.cn

Received: 15 October 2020; Accepted: 10 November 2020; Published: 13 November 2020

Abstract: Radial ultrasonic rolling electrochemical micromachining (RUR-EMM) is a new method of electrochemical machining (ECM). By feeding small and rotating electrodes aided by ultrasonic rolling, an array of pits can be manufactured, which is called microstructures. However, there still exists the problem of choosing the optimal machining parameters to realize the workpiece machining with high quality and high efficiency. In the present study, response surface methodology (RSM) was proposed to optimize the machining parameters. Firstly, the performance criteria of the RUR-EMM are measured through investigating the effect of working parameters, such as applied voltage, electrode rotation speed, pulse frequency and interelectrode gap (IEG), on material removal amount (MRA) and surface roughness (R_a). Then, the experimental results are statistically analyzed and modeled through RSM. The regression model adequacies are checked using the analysis of variance. Furthermore, the optimal combination of these parameters has been evaluated and verified by experiment to maximize MRA and minimize R_a. The results show that each parameter has a similar and non-linear influence on the MRA and R_a. Specifically, with the increase of each parameter, MRA increases first and decreases when the parameters reach a certain value. On the contrary, R_a decreases first and then increases. Under the combined effect of these parameters, the productivity is improved. The experimental value of MRA and R_a is 0.06006 mm^2 and 51.1 nm, which were 0.8% and 2.4% different from the predicted values.

Keywords: microstructure; radial ultrasonic rolling electrochemical micromachining (RUR-EMM); material removal amount; surface roughness; response surface methodology (RSM)

1. Introduction

The special properties of microstructures, such as drag reduction and noise reduction [1,2], material surface self-cleaning [3], self-repairing [4], antifriction [5], antifatigue [6] and improving load bearing [7], are expected to be widely used in engineering fields such as agricultural machinery, aerospace, mechanical engineering and so on. Generally, the processing material is the metal material that is difficult to process, and for its higher requirements about size, precision, surface quality and so on, result in its higher processing technology requirements. Compared with electrical discharge machining (EDM), laser machining, mechanical machining, etc. [8–10], electrochemical machining (ECM), which is widely used in the machining of metal surface microstructure [11–13], has the advantages of no loss of processing electrode, no residual stress on the surface after processing, no thermal influence layer, high machining surface quality, etc. With the deepening of research and the pursuit of high efficiency and precision, scholars try to add ultrasonic filed into ECM to explore the effect of multiphysical field machining. Ruszaj et al. [14] confirmed that the surface quality of the workpiece using ultrasonic electrochemical machining is better than pulsed electrochemical machining, and the addition of abrasive powder have a further improvement on the surface quality. Natsu et al. [15] verified that complex vibrations have a more advantageous effect on the replicating accuracy and processing speed than the longitudinal or lateral vibrations individually do in ECM.

The radial ultrasonic rolling electrochemical micromachining (RUR-EMM) combined rolling electrochemical micromachining (R-EMM) and ultrasonic vibration was studied [16], its machining efficiency and the quality of surface machined are both improved with the combination of their benefits. However, affected by many factors, such as applied voltage, pulse frequency, electrode rotation velocity, and the interelectrode gap, microstructure forming size is relatively difficult to control. Besides, due to the effect of ultrasonic, the interaction of each parameter is more complex. Therefore, it is necessary to find a suitable optimization algorithm and establish a theoretical model of the relationship between machining parameters and target parameters to find the best machining parameters.

Many scholars have explored the optimization of ECM. Based on analyzing the effect of parameters including workpiece, electrolyte and cathode, Shang et al. [17] developed a forward feed forward back propagation (BP) neural network to predict the anode accuracy in ECM. Combined BP neural network together with the Levenberg–Marquardt (L–M) optical algorithm, Zhu [18] proposed a digital cathode modification model to accurately design cathode and modify the turbine blade. Xu et al. [19] developed an electrochemical mechanical polishing (ECMP) prediction model on the basis of least squares support vector machines (SVMs) with the radial basis function and assessed the effect of polishing parameters including rotating speed, pressure of grinding tool, current density, grit size and machining time on surface roughness. Using response surface methodology (RSM), Munda et al. [20] established a model between predominant micromachining criteria, i.e., the material removal rate and the radial overcut and electrochemical processing parameters containing machining voltage pulse on/off ratio, machining voltage, electrolyte concentration, voltage frequency and tool vibration frequency and verified the accuracy of the model by ANOVA. Through RSM, Sen et al. [21] set up a significant model between important process parameters of electro jet drilling (EJD) such as applied voltage, capillary outside diameter, feed rate, electrolyte concentration and inlet electrolyte pressure and the quality of hole like roundness error, R_a and material removal rate (MRR) to improve the hole quality. Using RSM to set the cutting parameters such as the electrolyte concentration, electrolyte flowrate, applied voltage and tool feed rate as the variable and set material removal rate and surface roughness as the response, Senthilkumar et al. [22] optimize the ECM process parameters to maximize MRR and minimize R_a. Ei-Taweel et al. [23] developed a mathematic model to study the performance criteria of the electrochemical turning process through investigating the effect of working parameters, namely, applied voltage, wire feed rate, wire diameter, work piece rotational speed and overlap distance, on the metal removal rate, surface roughness and roundness error.

Based on the above research work, it is clear that the BP neural network, SVM and RSM are both optimal algorithm about the multiparameter. Notably, RSM could match the multiparameter to the multiresponse to analyze the relationship between variable and response. For RUR-EMM, it is necessary to explore the relationship of machining parameters, such as applied voltage, pulse frequency, electrode rotation velocity and interelectrode gap, and target parameters, such as material removal amount and surface roughness to get an optimal parameter, which is a multivariable and multiresponse problem. Therefore, response surface methodology is more applicable to RUR-EMM. In this study, an experiment using response surface methodology with four factors and five levels was designed to analyze and obtain the optimal machining parameters of RUR-EMM. The adequacy of the developed theoretical models was also tested by an analysis of variance (ANOVA) test.

2. Experiment Details

2.1. Experimental Setup

The proposed RUR-EMM system shown in Figure 1a comprises major subsystems: ultrasonic generator, pulsed power supply and rotary table controller, the electrolyte slot and rotary table, electrolyte supply, radial ultrasonic transducer and a numerical control laboratory mechanism with three, X, Y and Z, axes motions. As it shown in Figure 1b,c, the workpiece (anode) was fixed on the operating platform when it was processing, and the radial ultrasonic transducer with array micro bulge on the surface, which was clamped on the spindle of the machine tool, was used as the cathode. Under the control of numerical control system and rotation controller, the radial ultrasonic transducer rotated while feeding during machining and converted the electrical signal of ultrasonic generator into ultrasonic vibration. The electrolyte acted as a conductive medium and entered the electrolytic cell through the water pump, filling the entire interelectrode gap. When the pulse power is switched on, pits will be formed on the surface of the work piece due to electrochemical corrosion.

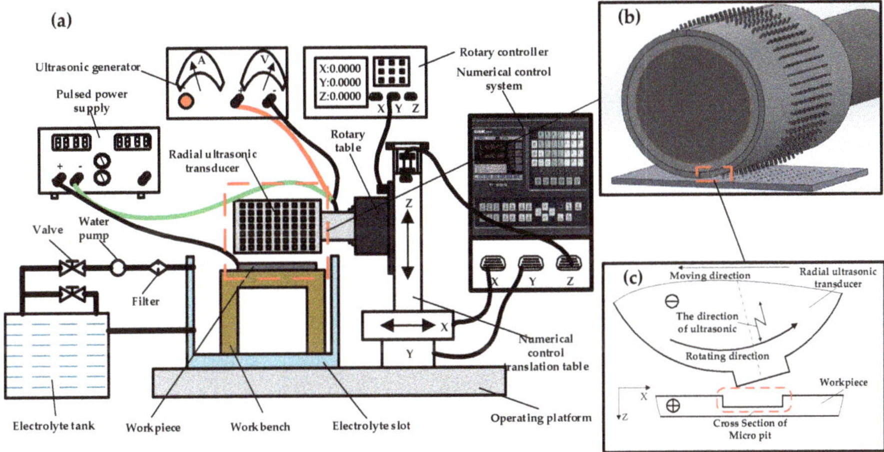

Figure 1. Schematic of (**a**) the proposed radial ultrasonic rolling electrochemical micromachining (RUR-EMM) system; (**b**) machining process and (**c**) machining principle.

2.2. Measurements Procedure

The material removal amount (MRA) of the sample is determined by the size of the cross-sectional area, which was measured using a 3D digital microscope (Keyence, Tokyo, Japan, VHX-7000). Additionally, then the area was calculated by establishing the coordinate and integrating the contour curve. The MRA was specified using the following equation:

$$MRA = S_{section} \tag{1}$$

The R_a of pit bottom for each specimen was measured by a white light interferometer (Chotest, Shenzheng, China, SuperView W1). 3D image was synthesized by the optical method, the bottom contour of the pit was extracted, and its roughness was measured. Each MRA and R_a's data was obtained by averaging three measurements. Based on these data, the following RSM optimization analysis was carried out.

3. Design of the Response Surface Test

3.1. Mathematical Model of Response Surface Methodology

The RSM approach is the procedure for determining the relationship between various process parameters with the various machining criteria and exploring the effect of these process parameters on the coupled responses, i.e., the MRA and R_a. Assuming that y is related to $x_1, x_2, \cdots, x_p$, set

$$Ey = f\left(x_1, \cdots, x_p\right) \tag{2}$$

where E_y is the response and $x_1, \cdots, x_p$ are the coded level of p quantitative variables. For the Equation (1) is unknown, so it is necessary to test (or sample), estimate Equation (1) from the test data obtained from a finite number of tests and use McLaughlin or Taylor expansion formula to fit, namely

$$f(x) \approx f(0) + \frac{f'(0)}{1!}x + \frac{f''(0)}{2!}x^2 + \tag{3}$$

Substitute Equation (2) into Equation (1)

$$Ey = f\left(x_1, \cdots, x_p\right) \approx a + bx_1 + \cdots + cx_p + \cdots + dx_p^2 + \cdots + ex_p^2 + fx_1x_2 + \cdots + gx_{p-1}x_p \tag{4}$$

In order to study the effect of ECM parameters on the two above-mentioned criteria, a second order polynomial response can be fitted into the following equation

$$Ey_u = a_0 + \sum_{i=1}^{p} a_i x_i + \sum_{i=1}^{p} a_{ii} x_i^2 + \sum_{i=1}^{p} a_{ij} x_i x_j \tag{5}$$

3.2. Experimental Design of Response Surface Machining

Response surface methodology is a collection of mathematical and statistical techniques useful for modeling and optimizing the response variable models involving quantitative independent variables. In order to achieve the accuracy and effectiveness of the experimental program, experiments were carried out according to a central composite design (CCD) based on RSM in this study [24,25].

Applied voltage, electrode rotation speed, pulse frequency and interelectrode gap are important factors affecting the quality of RUR-EMM. Based on the above factors, considering the actual processing conditions in Table 1. The experiments were designed at five levels as shown in Table 2. Table 3 presents the experimental design matrix and experimental results.

Table 1. Processing conditions.

Parameter	Value
Electrolyte concentration	10%NaNO$_3$
Electrolyte temperature (T_e)	25°
Protrusion size	70 mm × 50 mm
Ultrasonic vibration frequency(f_u)	28 KHz
Amplitude (A)	10 μm
Thickness of workpiece (T_w)	0.3 mm
Processing time	5 min
Power supply duty cycle	0.5
Workpiece material	SS 304

Table 2. Machining parameters and their levels.

Parameters	Units	Level				
		−2	−1	0	+1	+2
Applied voltage (U)	V	8	10	12	14	16
Electrode rotation speed (v)	°/s	0.05	0.15	0.1	0.2	0.25
Pulse frequency (f)	kHz	2	4	6	8	10
Inter-electrode gap (d)	μm	40	50	60	70	80

Table 3. Experimental results.

	Factors				Responses			Factors				Responses	
Run	U V	v °/s	f kHz	d μm	MRA mm^2	R_a of Pit Bottom nm	Run	U V	v °/s	f kHz	d μm	MRA mm^2	R_a of Pit Bottom nm
1	8	0.05	4	80	0.03025	116.7	17	12	0.15	6	80	0.05425	110.0
2	8	0.15	6	60	0.03634	179.5	18	12	0.15	10	60	0.05217	135.0
3	8	0.25	2	40	0.02826	200.5	19	12	0.25	6	60	0.05381	136.9
4	8	0.25	2	80	0.02814	209.9	20	14	0.1	4	50	0.05510	51.8
5	10	0.1	4	50	0.04321	112.9	21	14	0.1	4	70	0.05479	56.9
6	10	0.1	4	70	0.04290	116.6	22	14	0.1	8	50	0.05408	70.7
7	10	0.1	8	50	0.04291	130.4	23	14	0.1	8	70	0.05381	80.3
8	10	0.1	8	70	0.04173	139.3	24	14	0.2	4	50	0.05533	73.2
9	10	0.2	4	50	0.04337	133.7	25	14	0.2	4	70	0.05490	82.0
10	10	0.2	4	70	0.04302	140.7	26	14	0.2	8	50	0.05432	98.2
11	10	0.2	8	50	0.04241	160.0	27	14	0.2	8	70	0.05399	104.0
12	10	0.2	8	70	0.04212	171.9	28	16	0.05	10	80	0.03981	139.0
13	12	0.05	6	60	0.05337	102.1	29	16	0.15	6	60	0.04889	115.2
14	12	0.15	2	60	0.05345	105.6	30	16	0.25	2	80	0.04058	146.0
15	12	0.15	6	40	0.05439	109.7	31	16	0.25	10	40	0.03972	175.0
16	12	0.15	6	60	0.05649	107.6	-	-	-	-	-	-	-

3.3. Analysis of Response Surface Experimental Results

Using the collected results shown in Table 3, the final regression models for MRA and surface R_a as determined by the preceding analysis are

$$
\begin{aligned}
R_{MRA} = \quad & -0.14 + 0.020U + 0.15v + 4.560 \times 10^{-3}f + 1.293 \times 10^{-3}d \\
& + 4.438 \times 10^{-4}U{\cdot}v - 1.203 \times 10^{-5}U{\cdot}f \\
& + 2.969 \times 10^{-6}U{\cdot}d - 1.312 \times 10^{-4}v{\cdot}f + 6.875 \times 10^{-5}v{\cdot}d \\
& - 1.656 \times 10^{-6}f{\cdot}d - 7.003 \times 10^{-4}U^2 \\
& - 0.52v^2 - 3.748 \times 10^{-4}f^2 - 1.121 \times 10^{-5}d^2
\end{aligned}
\tag{6}
$$

$$
\begin{aligned}
R_{Ra} = \quad & 614.12 - 50.36U - 595.90v - 14.97f - 4.91d + 2.30U{\cdot}v - 0.35U{\cdot}f + 0.88U{\cdot}d + 17.76v{\cdot}f \\
& + 1.24v{\cdot}d - 9.081 \times 10^{-3}f{\cdot}d + 1.74U^2 + 1324.44v^2 + 1.63f^2 + 0.031d^2
\end{aligned}
\tag{7}
$$

The adequacy of the provided models was checked by the analysis of variance (as Tables 4 and 5). The significant and non-significant factors were tested through a Student's t test. Design Expert software (version 8.0.6, State Ease Inc., Minneapolis, MN, USA) was used to analyze the experimental data of the response parameters.

Table 4. Analysis of variance for the material removal amount (MRA).

Source	Sum of Squares	DF	Mean Square	F Value	Prob > F	
Model	1.121×10^{-3}	1	8.004×10^{-5}	19.68	<0.0001	significant
U	1.808×10^{-4}	1	1.808×10^{-4}	44.45	<0.0001	-
v	8.736×10^{-6}	1	8.736×10^{-6}	2.15	0.1622	-
f	1.307×10^{-5}	1	1.307×10^{-5}	3.21	0.0920	-
d	1.977×10^{-5}	1	1.977×10^{-5}	4.86	0.0425	-
$U{\cdot}v$	3.151×10^{-8}	1	3.151×10^{-8}	7.745×10^{-3}	0.9310	-
$U{\cdot}f$	3.706×10^{-8}	1	3.706×10^{-8}	9.109×10^{-3}	0.9251	-
$U{\cdot}d$	5.641×10^{-8}	1	5.641×10^{-8}	0.014	0.9077	-
$v{\cdot}f$	2.756×10^{-9}	1	2.756×10^{-9}	6.775×10^{-4}	0.9796	-
$v{\cdot}d$	1.891×10^{-8}	1	1.891×10^{-8}	4.647×10^{-4}	0.9465	-
$f{\cdot}d$	1.756×10^{-8}	1	1.756×10^{-8}	4.316×10^{-3}	0.9484	-
U^2	2.244×10^{-4}	1	2.244×10^{-4}	55.16	<0.0001	-
v^2	4.855×10^{-5}	1	4.855×10^{-5}	11.93	0.0033	-
f^2	6.426×10^{-5}	1	6.426×10^{-5}	15.80	0.0011	-
d^2	3.593×10^{-5}	1	3.593×10^{-6}	8.83	0.0090	-
Residual	6.509×10^{-5}	16	4.068×10^{-6}	-	-	-
Lack of fit	6.509×10^{-5}	10	6.509×10^{-6}	-	-	-
Pure Error	0.000	6	0.000	-	-	-
Cor Total	1.186×10^{-3}	30	-	-	-	-

Table 5. Analysis of variance for R_a.

Source	Sum of Squares	DF	Mean Square	F Value	Prob > F	
Model	15508	14	1107.72	6.02	0.0005	significant
U	1849.65	1	1849.65	10.06	0.0059	-
v	203.48	1	203.48	1.11	0.3085	-
f	209.42	1	209.42	1.14	0.3017	-
d	388.84	1	388.84	2.11	0.1652	-
$U{\cdot}v$	1.90	1	1.90	0.010	0.9203	-
$U{\cdot}f$	48.36	1	48.36	0.26	0.6151	-
$U{\cdot}d$	114.69	1	114.69	0.62	0.4412	-
$v{\cdot}f$	76.15	1	76.15	0.41	0.5290	-
$v{\cdot}d$	13.34	1	13.34	0.073	0.7911	-
$f{\cdot}d$	0.85	1	0.85	4.620×10^{-3}	0.9467	-
U^2	1714.33	1	1714.33	9.32	0.0076	-
v^2	386.14	1	386.14	2.10	0.1666	-
f^2	1458.79	1	1458.79	7.93	0.0124	-
d^2	339.94	1	339.94	1.85	0.1928	-
Residual	2942.13	16	183.88	-	-	-
Lack of fit	2942.13	10	204.44	-	-	-
Pure Error	0.000	6	0.000	-	-	-
Cor Total	18450.21	30	-	-	-	-

4. Results and Discussion

4.1. Effect of Machining Parameters on the Material Removal Amount

The effect of electrode rotation speed on the MRA is demonstrated in Figure 2a. Obviously, the MRA increased first and then decreased with the increase of the electrode rotation speed. The MRA reached a maximum of 0.0575 mm^2 when the electrode rotation speed was 0.15°/s. The reason may be that when the applied voltage was stable and the rotating speed was low (less than 0.15°/s), the material exchange in the gap flow field was slow, and this suppressed the material removal on the workpiece surface. However, with the rotating speed increasing, the material exchange in the gap flow field became faster, the electrical conductivity increased in a certain extent, the current density increased and the material removal amount became larger. When the rotational speed was faster than 0.15°/s, the machining time for the workpiece surface deceased and less material will be eroded. Figure 2b shows the effect of applied voltage on MRA. It is obvious that the applied voltage influenced the MRA nonlinearly. Notably, the MRA increased with the increase of the applied voltage until the applied voltage reached a certain value of 14 V, meanwhile, the MRA reached its peak of 0.0596 mm^2. Then, it decreased. This may be due to the change of electric field. Under the condition of stable flow field, the gap material exchange in electrochemical machining was relatively stable (the increase rate of byproducts in anodic dissolution field was less than that of fluid), resulting in relatively constant conductivity. When the applied voltage was relatively low (less than 14 V), with the increase of the voltage, the gap electric field intensity increased, and with the constant conductivity, the gap current density also increased, leading to a corresponding increase in the material removal amount. When the processing voltage was greater than 14 V, the product increase rate of anodic dissolution field was greater than that of the fluid, and the gap concentration polarization increased, hindering the normal anodic dissolution, resulting in the reduction of the material removal amount.

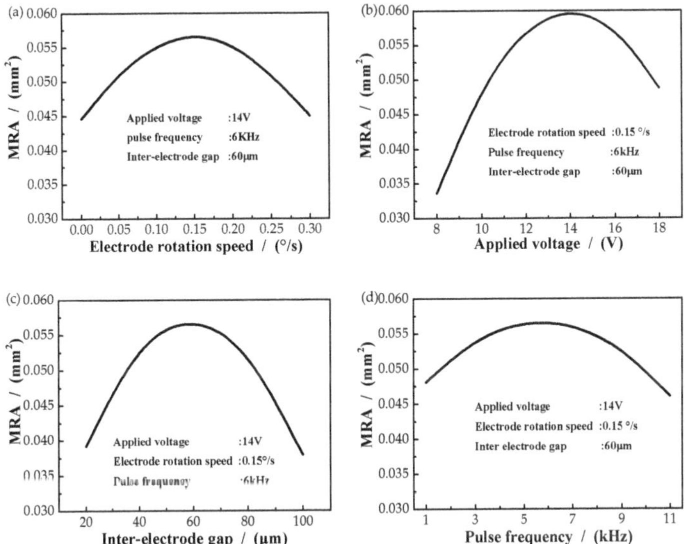

Figure 2. Main effects of process parameters on MRA. (**a**) Electrode rotation speed; (**b**) Applied voltage; (c)Interelectrode gap; (**d**)Pulse frequency.

The effect of the interelectrode gap (IEG) on MRA is shown in Figure 2c. It is noted that the MRA increased with the increase of IEG until it got its peak of 0.555 mm^2, and then it decreased. This is because under the condition of stable applied voltage, when the IEG was small (less than 60 um), the electric field intensity between electrodes gradually decreased with the increase of the IEG. However, the increased gap made the material exchange of the gap flow field faster and the gap concentration polarization smaller, which was conducive to the dissolution of anode products and led to the increase of material removal amount. However, when the IEG was greater than 60 μm, the material exchange in the clearance flow field was relatively stable and the electrical conductivity was relatively stable. The increased clearance reduced the current density and led to a corresponding decrease in the material removal amount. Figure 2d revealed the influence of pulse frequency of the MRA. It is noted that the MRA increased with the increase of pulse frequency until it got its peak of 0.552 mm^2, and then it decreased. The reason may be that when the applied voltage is stable and the pulse frequency is low (below 6 kHz), with the increase of pulse frequency, the anodic dissolution time becomes shorter, which is conducive to the material exchange in the gap flow field, leading to the increase of electrical conductivity, current density and material removal amount. However, when the pulse frequency is larger than 6 kHz, the anodic dissolution time becomes less and less, so that the material removal amount decreases.

To sum up, each parameter will have an impact on MRA. When applied voltage, electrode rotation speed, pulse frequency and interelectrode gap changed, the value of MRA always increased first and then decreased. Among them, applied voltage had the greatest influence on MRA. When applied voltage changed, the change value of MRA reached 0.0275 mm^2. Figure 3 shows the interactions between process parameters on MRA. The curved surface of MRA was like a 'convex hull', there was a maximum at the peak point under the interaction of parameters.

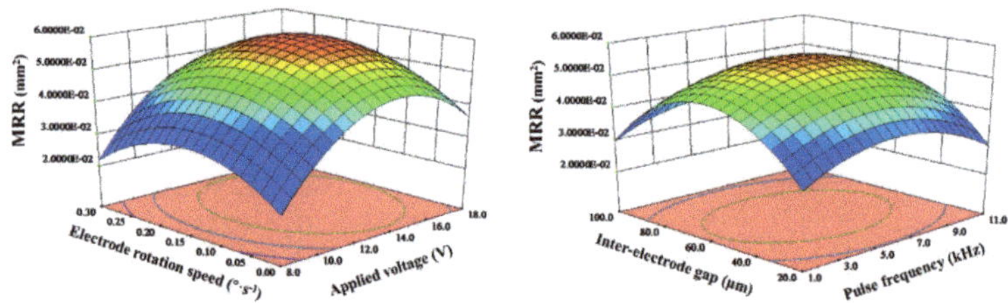

Figure 3. The interactions between process parameters on MRA.

4.2. Effect of Machining Parameters on Surface Roughness

Surface roughness (R_a) can display the quality of a machined surface. It is necessary to investigate the effects of parameters on R_a. Figure 4a shows the influence of electrode rotation speed on R_a. R_a decreased from 83 to 55 nm when the electrode rotation speed increased from 0 to 0.15°/s, and then it increased from 55 to about 84 nm when electrode rotation speed increased from 0.15 to 0.5°/s. This is because under the condition of stable applied voltage, at a low speed (less than 0.15°/s), with the increase of speed, the material exchange in the gap flow field became faster, the conductivity increased to a certain extent, the current density increased and the surface roughness decreased. When the rotational speed was faster than 0.15°/s, the surface processing time of the material was reduced to less anodic dissolution. Coupled with rotation, the electric field between poles was not uniform, resulting in worse surface roughness. The effect of applied voltage on R_a is shown in Figure 4b. R_a decreased from 104 to 47 nm when applied voltage increased from 8 to 14 V, and then it increased from 47 to about 90 nm when applied voltage increased from 14 to 16 V. A reason may be given that under the condition of a stable flow field, the gap material exchange in electrochemical machining was relatively stable (the increase rate of products in anodic dissolution field was less than that of fluid), resulting in relatively constant gap conductivity. When the applied voltage was relatively low (less than 14 V), with the increase of the voltage, the gap electric field intensity increased, and with the constant conductivity, the gap current density also increased, leading to a corresponding increase in the material removal amount. When the processing voltage was greater than 14 V, the product of anodic dissolution field increased faster than the fluid velocity, and the gap electric field was not evenly distributed, resulting in the decrease of surface roughness.

The effect of the interelectrode gap (IEG) is illustrated in Figure 4c. R_a decreased from 104 to 50 nm when IEG increased from 40 to 60 μm, and then it increased from 50 to about 106 nm when IEG increased from 60 to 80 μm. This result can be explained that under the condition of stable applied voltage, when the processing gap was small (less than 60 um), the electric field intensity between electrodes gradually decreased with the increase of the IEG. However, the increased clearance makes the material exchange of the clearance flow field faster and the gap concentration polarization smaller, which was conducive to the dissolution of anode products and led to the increase of surface roughness. However, when the machining gap was greater than 60 μm, the material exchange in the gap flow field was relatively stable and the electrical conductivity was relatively stable. The increased IEG reduced the current density and led to a corresponding decrease in surface roughness. Figure 4d revealed the effect of pulse frequency on R_a. R_a decreased from 84 to 47 nm when pulse frequency increased from 2 to 6 kHz, and then it increased from 47 to about 110 nm when pulse frequency increased from 60 to 80 μm. It can be explained that when the applied voltage was stable and the pulse frequency was low (less than 6 kHz), with the increase of the pulse frequency, the anodic dissolution time became shorter, which is conducive to the material exchange of the clearance flow field, the uniform distribution of the electric field between poles, and the decrease of the surface roughness. However, when the pulse frequency was larger than 6 kHz, the anodic dissolution time became less and less, so that the surface roughness increased.

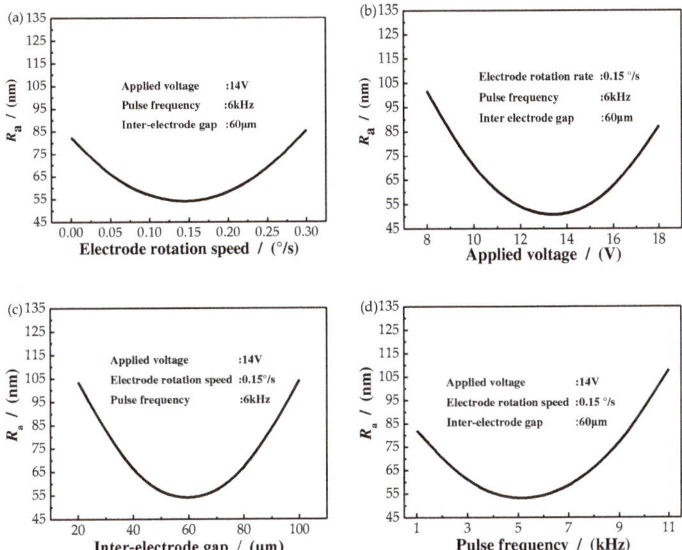

Figure 4. Main effects of process parameters on R_a. (**a**)Electrode rotation speed; (**b**) Applied voltage; (**c**) Interelectrode gap; (**d**) Pulse frequency.

Above all, each parameter will have an impact on R_a. When applied voltage, electrode rotation speed, pulse frequency and interelectrode gap varied, the value of R_a always decreased first and then increased. Among them, applied voltage had the greatest influence on R_a. When applied voltage changed, the change value of MRA reached 57 nm. Figure 5 shows the interactions between process parameters on R_a. The curved surface of R_a was like a 'valley', R_a would reach its minimum at the bottom point under the interaction of parameters.

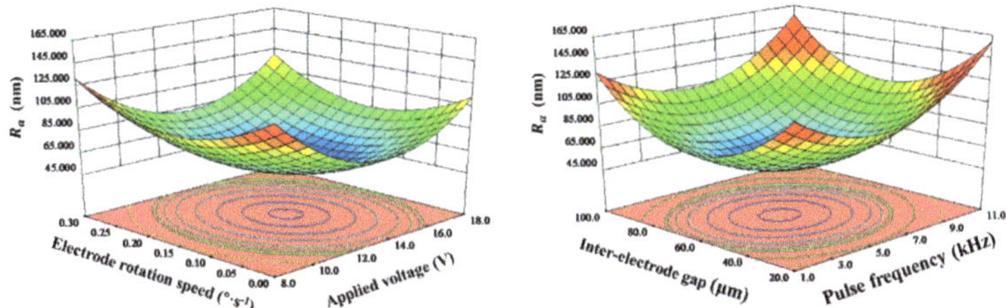

Figure 5. The interactions between process parameters on R_a.

4.3. Multiresponse Optimization of the Process

Selection of the optimal machining parameter combination for achieving improved process performance, e.g., material removal amount and surface roughness, is a challenging task in RUR-EMM due to the presence of a large number of process variables and complicated stochastic process mechanisms. Based on the experimental results data shown in Sections 4.1 and 4.2, it could be seen there existed the optimal parameters for the maximum MRA and the minimum R_a of the machined workpiece surface.

At the same time, the distribution diagram of the optimal machining parameters for MRA and R_a was analyzed. Figure 6 shows the distribution of optimal machining parameters with maximum MRA, where applied voltage (U) was 14.08 V, electrode rotation velocity (v) was 0.15°/s, pulse frequency(f) was 5.70 kHz, interelectrode gap (d) was 59.54 μm and corresponding MRA was 0.0596 mm². Figure 7 shows the distribution of optimal machining parameters with the minimum R_a, where applied voltage (U) was 13.40 V, electrode rotation velocity (v) was 0.15°/s, pulse frequency (f) was 5.35 kHz, interelectrode gap (d) was 57.76 μm and the corresponding R_a was 49.895 nm. Therefore, the average value from each parameter was approximately seen as an optimal parameter shown in Table 6.

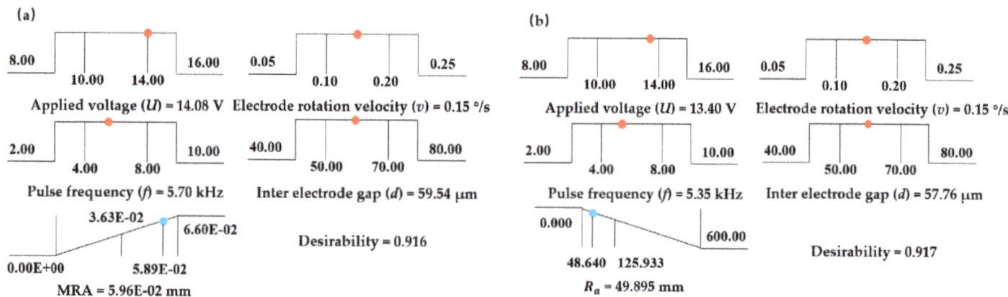

Figure 6. Optimal machining parameters with (**a**) maximum MRA and (**b**) minimum R_a.

Table 6. Optimum value of parameters.

Parameter	Optimum Value
Applied voltage (V)	14.7
Electrode rotation speed (°/s)	0.15
Pulse frequency (kHz)	5.5
Inter-electrode gap (μm)	58.6

A verification experiment was performed using the optimized parameters shown in Table 4. After the experiment of the optimal processing parameters, the MRA of the micro dimple was 0.06006 mm² and R_a was 51.1 nm, which were 0.8% and 2.4% different from the predicted values, respectively. Figure 7 shows the cross sectional area of the micro pit and its R_a. Clearly, the value of MRA was greater than the maximum MRA (0.05649 mm²) without optimization, and the value of R_a was less than the minimum R_a (51.8 nm) without optimization, indicating that the optimization model was effective and successful.

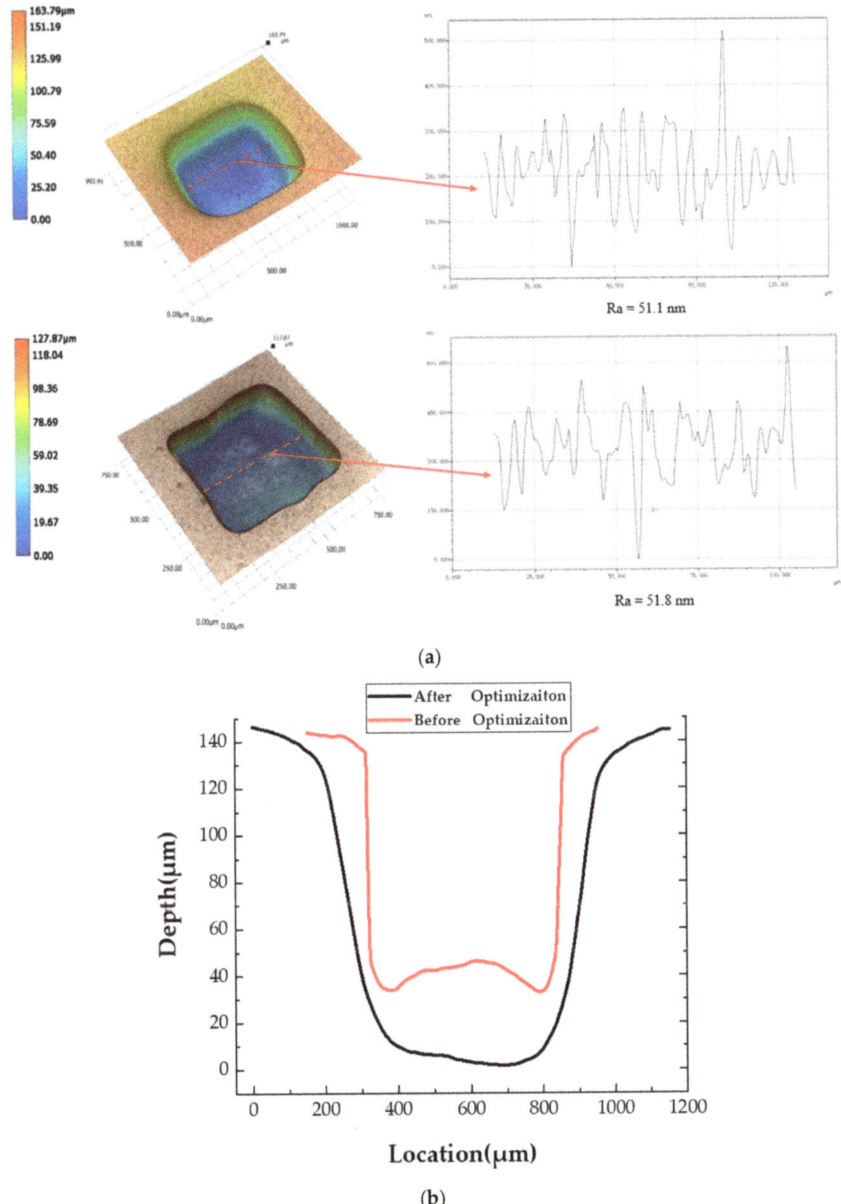

Figure 7. The R_a and MRA before and after optimization. (**a**) The morphology of micro pit and its R_a; (**b**) The cross section of micro pit.

5. Conclusions

Based on the experimental method of the Central Composite Design, 31 experimental studies using radial ultrasonic rolling electrochemical micromachining (RUR-EMM) under different working conditions were carried out, and according to the response surface optimization analysis, global optimization was carried out by using the Design Expert software, and the combination of optimization parameters in a certain range was found. The following conclusion can be drawn:

(1) Response surface methodology is a suitable data optimization algorithm to radial ultrasonic rolling electrochemical micromachining.

(2) Parameters, including applied voltage, electrode rotation speed, pulse frequency and the interelectrode gap all had a nonlinear effect on the MRA and R_a. Especially, the applied voltage has.

(3) The optimum combination of parameters of applied voltage 14.70 V, electrode rotation speed 0.15°/s, pulse frequency 5.5 kHz, interelectrode gap 58.6 μm for maximizing the metal removal rate of 0.06006 mm^2 and a minimizing surface roughness of 51.1 nm could be obtained.

Author Contributions: Conceptualization, K.H. and M.W.; methodology, K.H. and X.C.; software, X.C.; validation, K.H., X.C. and M.W.; formal analysis, K.H.; investigation, X.C.; resources, K.H.; data curation, K.H.; writing—original draft preparation, K.H.; writing—review and editing, M.W.; visualization, X.C.; supervision, M.W.; project administration, M.W.; funding acquisition, M.W. All authors have read and agreed to the published version of the manuscript.

Funding: This research was funded by [National Natural Science Foundation of China] grant numbers [51975532] And [Natural Science Foundation of Zhejiang Province] grant numbers [LY19E050007].

Conflicts of Interest: The authors declare no conflict of interest.

References

1. Walsh, M.J.; Michael, J. Riblets as a viscous drag reduction technique. *AIAA J.* **1983**, *21*, 485–486. [CrossRef]

2. Yibin, L.; Huimin, G.; Yu, J. Design and preparation of biomimetic polydimethylsiloxane (PDMS) films with superhydrophobic, self-healing and drag reduction properties via replication of shark skin and SI-ATRP. *Chem. Eng. J.* **2018**, *356*, 318–328.

3. Vorobyev, A.Y.; Guo, C. Multifunctional surfaces produced by femtosecond laser pulses. *J. Appl. Phys.* **2015**, *117*, 033103.1–033103.5. [CrossRef]

4. Riveiro, A.; Abalde, T. Influence of laser texturing on the wettability of PTFE. *Appl. Surf. Sci.* **2020**, *515*, 145984–145993. [CrossRef]

5. Xie, F.; Lei, X.B. The Influence of Bionic Micro-Texture's Surface on Tool's Cutting Performance. *Key Eng. Mater.* **2016**, *693*, 1155–1162. [CrossRef]

6. Wu, L.Y.; Lv, Y.; Song, Y.Q. Experimental Investigation on Anti-Fatigue Function of Gear Surface with Grid Micro-Morphology. *Adv. Mater. Res.* **2013**, *652–654*, 1842–1845. [CrossRef]

7. Chen, L.Y.; Li, R.; Xie, F. Load-bearing capacity research in wet clutches with surface texture. *Measurement* **2019**, *142*, 96–104. [CrossRef]

8. Bialo, D.; Peronczyk, J. Electrodischarge drilling of microholes in aluminium matrix composites. *Int. J. Mach. Mach. Mater.* **2018**, *3*, 272–282. [CrossRef]

9. Wang, M.; Liu, Q.; Zhang, H. Laser direct writing of tree-shaped hierarchical cones on a superhydrophobic film for high efficiency water collection. *ACS Appl. Mater. Interfaces* **2017**, *9*, 29248–29254. [CrossRef]

10. Wyszynski, D.; Bizon, W.; Miernik, K. Electrodischarge Drilling of Microholes in c-BN. *Micromachines* **2020**, *11*, 179. [CrossRef]

11. Thanigaivelan, R.; Arunachalam, R.M.; Karthikeyan, B.; Loganathan, P. Electrochemical micromachining of stainless steel with acidified sodium nitrate electrolyte. *Proc. CIRP* **2013**, *6*, 351–355. [CrossRef]

12. Das, A.K.; Kumar, P.; Sethi, A. Influence of process parameters on the surface integrity of micro holes of SS304 obtained by micro-EDM. *J. Braz. Soc. Mech. Sci. Eng.* **2016**, *38*, 2029. [CrossRef]

13. Wang, M.; Bao, Z.; Qiu, G. Fabrication of micro-dimple arrays by AS-EMM and EMM. *Int. J. Adv. Manuf. Technol.* **2017**, *93*, 787–797. [CrossRef]

14. Ruszaj, A.; Zybura, M.; Zurek, R. Some aspects of the electrochemical machining process supported by electrode ultrasonic vibrations optimization. *Proc. Inst. Mech. Eng. Part B J. Eng. Manuf.* **2003**, *217*, 1365–1371. [CrossRef]

15. Natsu, W.; Nakayama, H.; Yu, Z. Improvement of ECM characteristics by applying ultrasonic vibration. *Int. J. Precis. Eng. Manuf.* **2012**, *13*, 1131–1136. [CrossRef]

16. Wang, M.; Chen, X. Influences of gap pressure on machining performance in radial ultrasonic rolling electrochemical micromachining. *Int. J. Adv. Manuf. Technol.* **2020**, *107*, 157–166. [CrossRef]

17. Shang, G.Q.; Sun, C.H. Application of BP Neural Network for Predicting Anode Accuracy in ECM. In Proceedings of the IEEE 2008 International Symposium on Information Science and Engineering (ISISE), Shanghai, China, 20–22 December 2008; pp. 428–432.

18. Dong, Z. Experimental Study on the Cathode Digital Modification of Turbine Blade in Electrochemical Machining. *J. Mech. Eng.* **2011**, *47*, 191–198.

19. Xu, W.; Wei, Z.; Sun, J. Surface quality prediction and processing parameter determination in electrochemical mechanical polishing of bearing rollers. *Int. J. Adv. Manuf. Technol.* **2012**, *63*, 129–136. [CrossRef]

20. Munda, J.; Bhattacharyya, B. Investigation into electrochemical micromachining (EMM) through response surface methodology based approach. *Int. J. Adv. Manuf. Technol.* **2008**, *35*, 821–832. [CrossRef]

21. Sen, M.; Shan, H.S. Analysis of hole quality characteristics in the electro jet drilling process. *Int. J. Mach. Tools Manuf.* **2005**, *45*, 1706–1716. [CrossRef]

22. Senthilkumar, C.; Ganesan, G.; Karthikeyan, R. Bi-performance optimization of electrochemical machining characteristics of Al/20%SiCp composites using NSGA-II. *Proc. Inst. Mech. Eng. Part B J. Eng. Manuf.* **2010**, *224*, 1399–1407. [CrossRef]

23. El-Taweel, T.A.; Gouda, S.A. Performance analysis of wire electrochemical turning process—RSM approach. *Int. J. Adv. Manuf. Technol.* **2011**, *53*, 181–190. [CrossRef]

24. Espejo, M.R. Design of Experiments for Engineers and Scientists. *Technometrics* **2006**, *48*, 304–305. [CrossRef]

25. Ebeid, S.J.; Hewidy, M.S.; El-Taweel, T.A.; Youssef, A.H. Towards higher accuracy for ECM hybridized with low frequency vibrations using the response surface methodology technique. *J. Mater. Process Technol.* **2004**, *149*, 428–434. [CrossRef]

Publisher's Note: MDPI stays neutral with regard to jurisdictional claims in published maps and institutional affiliations.

 *micromachines*

Article

A 2D Waveguide Method for Lithography Simulation of Thick SU-8 Photoresist

Zi-Chen Geng, Zai-Fa Zhou * , Hui Dai and Qing-An Huang

Key Laboratory of MEMS of the Ministry of Education, Southeast University, Nanjing 210096, China; 220184753@seu.edu.cn (Z.-C.G.); 220184865@seu.edu.cn (H.D.); hqa@seu.edu.cn (Q.-A.H.)
* Correspondence: zfzhou@seu.edu.cn; Tel.: +86-025-8379-2632 (ext. 8817)

Received: 30 September 2020; Accepted: 27 October 2020; Published: 29 October 2020

Abstract: Due to the increasing complexity of microelectromechanical system (MEMS) devices, the accuracy and precision of two-dimensional microstructures of SU-8 negative thick photoresist have drawn more attention with the rapid development of UV lithography technology. This paper presents a high-precision lithography simulation model for thick SU-8 photoresist based on waveguide method to calculate light intensity in the photoresist and predict the profiles of developed SU-8 structures in two dimension. This method is based on rigorous electromagnetic field theory. The parameters that have significant influence on profile quality were studied. Using this model, the light intensity distribution was calculated, and the final resist morphology corresponding to the simulation results was examined. A series of simulations and experiments were conducted to verify the validity of the model. The simulation results were found to be in good agreement with the experimental results, and the simulation system demonstrated high accuracy and efficiency, with complex cases being efficiently handled.

Keywords: lithography simulation; microelectromechanical system; waveguide method

1. Introduction

1.1. Research Background

The lithography process refers to the process of transferring geometric pattern on the mask to the photoresist coated on the surface of the semiconductor wafer. With the rapid development of semiconductor technology, the lithography process has become an important propulsive force to achieve integrated and miniaturized devices. SU-8 photoresist [1] is a chemically amplified negative-tone resist based on epoxy resin, which has excellent chemical stability, resistance against solvents, and mechanical and optical properties [2]. It has become a mainstream micromachining technology for high aspect ratio structures in the field of microelectromechanical systems (MEMS) [3], including micrototal analysis systems [4], microfluidic systems [5], electroplating [6], and so on.

The propagation process of light during exposure is shown in Figure 1. First, we assume that the incident light is parallel, uniform, and perpendicular to the mask surface. The incident light is diffracted when passing through the mask hole. Then, when the light reaches the interface between the air and the photoresist, part of the light will be refracted and will continue to propagate through the photoresist, and the other part will be reflected back into the air. Part of the light that continues to propagate to the inside of the photoresist will be absorbed by the photoresist and generate photoacid at the same time. Finally, at the photoresist/substrate interface, part of the light is refracted into the substrate, and part of the light is reflected back into the photoresist and superimposed with the incident light. In general, the entire propagation process can be regarded as the propagation of diffracted light, taking into account other optical phenomena. Among them, the diffraction effect has the most important influence on light intensity distribution in the photoresist.

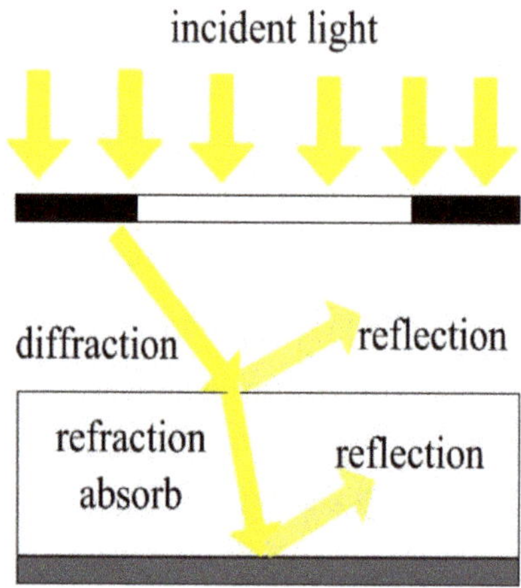

Figure 1. Light propagation process.

1.2. Research Significance

Because of expensive equipment and complicated process steps in the lithography process, we need repeated experiments, which is costly and time consuming. It is difficult to understand the inherent principle of lithography technology. Therefore, lithography simulation technology [7] has become a useful tool for predicting and optimizing the manufacturing process, which is also an integral part of lithography technology. Lithography simulation technology avoids the problems of high cost and time consumption caused by repeated plate making and tape production. At the same time, it can optimize the process parameters and promote the development of lithography technology. The flow chart of lithography simulation is as follows (Figure 2).

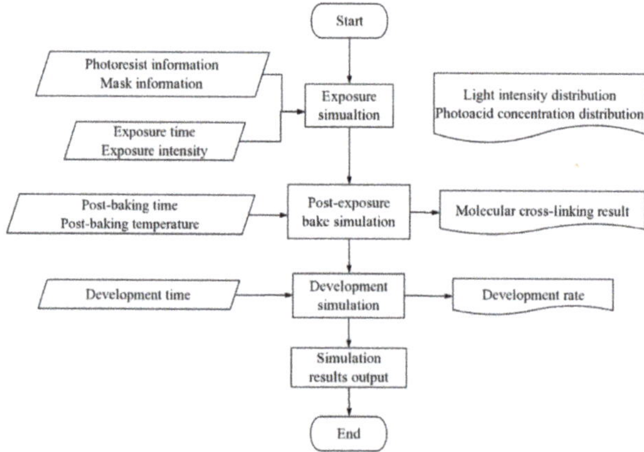

Figure 2. Flow chart of lithography simulation.

The final morphology of thick SU-8 photoresist after development depends largely on the exposure process [8]. The developed photoresist morphology can be predicted by simulating light intensity distribution in the photoresist. Therefore, it has become an inevitable and useful requirement to establish a lithography simulation model [9] of SU-8 thick photoresist to calculate light intensity distribution.

In this paper, a simulation model for UV lithography of SU-8 negative thick photoresist is presented to predict the profiles of developed SU-8 structures based on rigorous electromagnetic field theory. The model schematic is shown in Figure 3. We used the light intensity distribution calculation model to calculate the distribution of light intensity based on mask information, light source information, and photoresist information. The parameters of the air gap, incident angle, and photoresist depth, which have significant influence on the profile quality of developed structures, were studied in the simulation. Simulation results based on profiles of exposure intensity distribution were compared with the experimental results to validate the model.

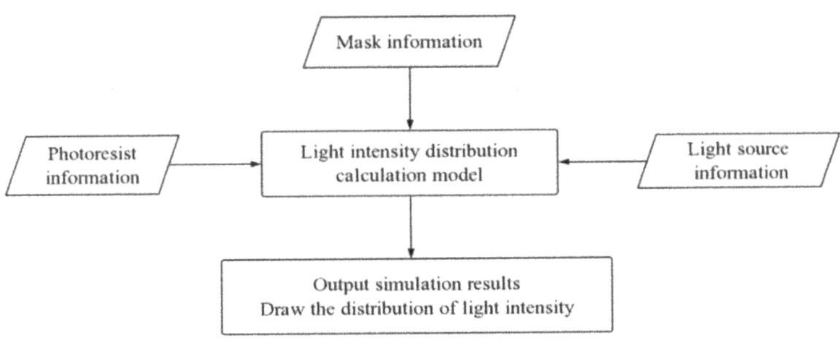

Figure 3. Model schematic.

2. Materials and Methods

Light intensity simulations are based on two main categories of models. One category is based on scalar diffractive theory [10], and the other is based on rigorous electromagnetic field theory [11], including the finite difference time domain method (FDTD) and the waveguide (WG) method. The limitation of these models comes from simulation accuracy, application capability for special lithography methods such as inclined SU-8 photoresist lithography, and numerical costs for large-scale 3D simulations. Therefore, considering calculation accuracy and calculation time, we finally used the waveguide method to calculate light intensity distribution.

The waveguide [12,13] method has been developed and employed in lithography simulations during the past decade. The idea of the waveguide method is based on rigorous electromagnetic field theory, and it is well adapted to lithography simulation with many typical mask geometries. The first prototype of the waveguide method was presented by Nyyssonen [14] in order to simulate the measurement of optical linewidth. Then, this method was used to simulate diffraction when light passes through a 2D-phase shift mask based on Yuan's theory [15] and later extended to 3D domain [16] by Lucas.

The model is based on the 2D spatial frequency solution to Maxwell's equations, the so-called waveguide method [17–21]. Firstly, the simulated structure is sliced into different rectangular layers with homogeneous optical properties in the Z-axis direction, as shown in Figure 4. Then, the material parameters and electromagnetic fields in each layer are expanded by Fourier series. Simultaneous equations formed by these expanded series and Maxwell equations give an eigenvalue problem. After that, a hybrid approach using vector potentials is applied to couple electromagnetic field in each layer. Finally, each layer is coupled according to the proper boundary continuity condition, and the electromagnetic field information in the photoresist is converted into light intensity information.

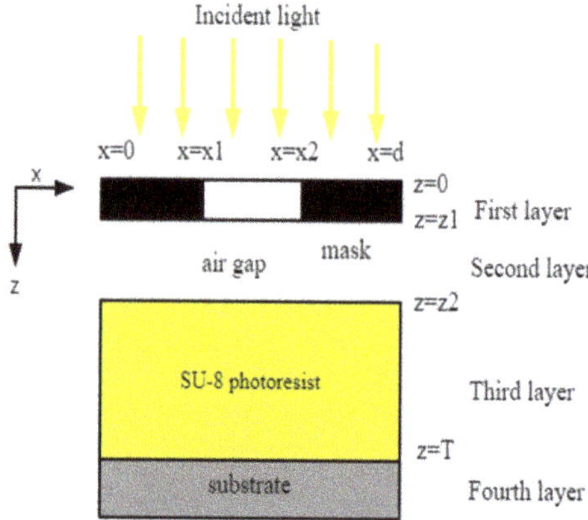

Figure 4. The two-dimensional schematic of the waveguide method.

Firstly, the coupled Maxwell's equation determined by vector potential [22] can be simplified to two differential equations by separating the variable; the equation contains the complex potential $k^2\varepsilon$ and the eigenvalue α^2:

$$\frac{d^2X}{dx^2} + k^2\varepsilon X = \alpha^2 X$$
$$\frac{d^2Z}{dz^2} = -\alpha^2 Z \tag{1}$$

Before solving the equation numerically, the dielectric constant ε needs to be expanded by Fourier series. It is assumed that the mask structure is arranged infinitely and repeatedly, with the length d as the period in the x-direction:

$$\varepsilon^j(x) = \sum_{q=-L}^{+L} \varepsilon_q^j \exp(i2\pi qbx) \tag{2}$$

The coefficient of each term can be obtained by inverse Fourier transform:

$$\varepsilon_q^j = b\int_0^d \varepsilon^j(x)\exp(-i2\pi qbx)dx$$
$$X(x) = \sum_l B_l \exp(i2\pi lx/d) \tag{3}$$

Then, the material parameters and electromagnetic fields in each layer are expanded by Fourier series. Simultaneous equations formed by these expanded series and Maxwell equations give an eigenvalue problem:

$$DB = \alpha^2 B$$
$$D_{pq} = k^2\varepsilon_{p-q} - (2\pi p/d)^2\delta_{p,q} \tag{4}$$

where α and B are the eigenvalues and eigenvectors of a (2L + 1) by (2L + 1) matrix D. The computation time depends on the number of orders of the Fourier series expansion and the number of rectangular layers.

The electromagnetic field distribution in layer j is as follows. α_m^j, B_m^j is obtained by solving the eigenvalue problem, and A_m^j, A'_m^j is obtained using the boundary continuity condition:

$$\begin{cases} E_y^j = \sum_m [A_m^j \exp(i\,\alpha_m^j(z-z_j)) + A'_m^j \exp(-i\,\alpha_m^j(z-z_j))] \times \sum_l B_{l,m}^j \exp(i2\pi lx/d) \\ H_x^j = \frac{i}{k}\frac{\partial E_y^j}{\partial z} \qquad H_z^j = -\frac{i}{k}\frac{\partial E_y^j}{\partial x} \end{cases} \tag{5}$$

After making this transformation and solving the eigenvalue problem, the boundary continuity conditions are matched so that the tangential components of the electric and magnetic fields have to be continuous at $z = 0$ [23]. The boundary continuity condition is as follows:

$$E_y^j = E_y^{j+1} \quad H_y^j = H_y^{j+1} \tag{6}$$

In the air and first layer interface:

$$\begin{bmatrix} C_{11}^0 & C_{12}^0 \\ 0 & 0 \end{bmatrix} \begin{bmatrix} A^1 \\ A'^1 \end{bmatrix} = \begin{bmatrix} R \\ 0 \end{bmatrix} \tag{7}$$

$$\begin{aligned} R_l &= 2[1 - (l_0 b \lambda_0)^2]^{\frac{1}{2}} \delta_{l,l_0} \\ &= 2[1 - (\sin \theta)^2]^{\frac{1}{2}} \delta_{l,l_0} \end{aligned} \tag{8}$$

where θ is the incident angle, and l_0 is the order of incident plane wave. It is convenient for us to change the different parameters to obtain the corresponding results.

By matching the boundary conditions at $z = T$, we obtain the following:

$$\begin{bmatrix} C_{11}^n & C_{12}^n \\ 0 & 0 \end{bmatrix} \begin{bmatrix} A^n \\ A'^n \end{bmatrix} = \begin{bmatrix} 0 \\ 0 \end{bmatrix} \tag{9}$$

At an intermediate interface $z = z_j$, the waves in layer $j+1$ are related to the waves in layer j by

$$\begin{bmatrix} C_{11}^j & C_{12}^j \\ C_{21}^j & C_{22}^j \end{bmatrix} \begin{bmatrix} A^j \\ A'^j \end{bmatrix} = \begin{bmatrix} E_{11}^{j+1} & E_{12}^{j+1} \\ E_{21}^{j+1} & E_{22}^{j+1} \end{bmatrix} \begin{bmatrix} A^{j+1} \\ A'^{j+1} \end{bmatrix} \tag{10}$$

where matrix C contains only the topography information, such as mask size, material parameters, and so on, and the vector $[A, A']$ contains diffraction information.

Finally, the light intensity can be calculated by the electric field value:

$$I(p, q) = n_r \left| E_y^j(p, q) \right|^2 \tag{11}$$

The incident angle is due to refraction at the air/photoresist interface. The structural inclined angle and the structural width of the developed photoresist vary with the incident angle. The relationship between the incident angle δ, the refractive angle θ, and the structural inclined angle α is determined by Snell's law as follows:

$$\Theta = \sin^{-1}(n_1 \cdot \sin \delta / n_2) \tag{12}$$

$$\alpha = 90 - \theta \tag{13}$$

where $n_1 (= 1)$ and $n_2 (= 1.67)$ are the refractive indices of the air and SU-8, respectively.
More details of the method is described in reference [24–26].

3. Results and Discussion

3.1. Simulation Results

The simulation results showed the distribution of normalized intensity calculated under different conditions. In addition to finding light intensity distribution on a certain layer of photoresist, a contour map can also describe the intensity value of a two-dimensional profile of photoresist. This method can help intuitively predict the final development morphology of the photoresist. Observing the influence of different parameters on light intensity distribution provides a guide to set the corresponding process parameters so as to optimize the structure design.

In the exposure process, the surface photolithographic dose D (mJ/cm^2) is multiplied by the exposure time and the incident light intensity. The surface exposure dose can be considered to be constant over the entire thickness of the photoresist layer. Photolithography of thick-film photoresist implies that the thickness of the photoresist material needs to be taken into account in order to determine the optimal exposure dose for a given photoresist film thickness. The exposure dose is the light intensity at a given thickness and exposure time. According to the material parameters of the photoresist and substrate, the depth of the photoresist can be calculated to obtain the optimal exposure dose under the given photoresist depth. In this calculation, the incident light intensity was 2.6 mW/cm^2, its wavelength was 365 nm, and the thickness of SU-8 photoresist was 300 μm. According to our simulation conditions, when the thickness of SU-8 photoresist is 300 μm, the optimal exposure dose is 140 (mJ/cm^2) [27].

First, we analyzed the simulation of the intensity of light at vertical incidence. The intensity curves for vertical exposure with different thicknesses are shown in Figure 5, and the corresponding contour maps are shown in Figure 6. We can see from the results that the intensity in SU-8 was simultaneously attenuated along the radiation direction with increasing photoresist thickness because of the diffraction and absorption of the incident light in the photoresist.

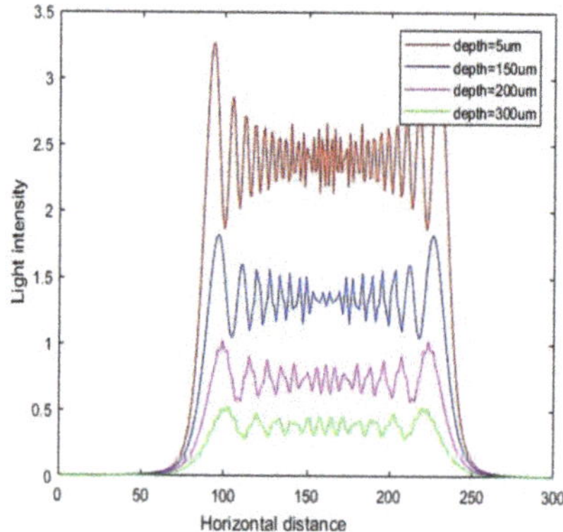

Figure 5. The intensity curve for vertical exposure with different depths.

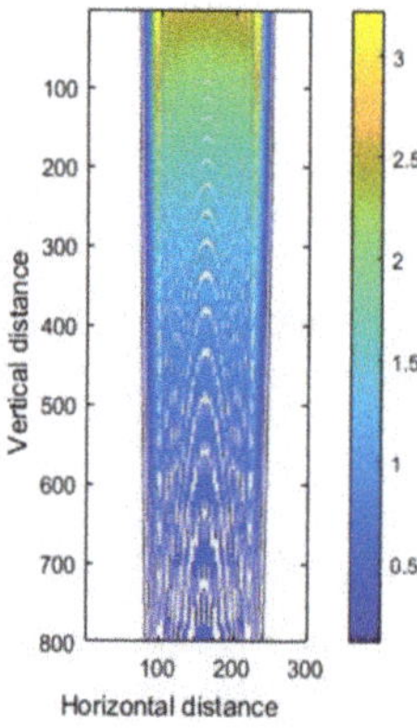

Figure 6. The contour map of intensity distribution for vertical exposure.

The previous simulation was based on the case of a single mask hole. In practice, there are often multiple mask holes, and selection of the spacing between the mask holes is an important parameter to be investigated when designing the layout. Different mask hole spacing will affect the final development morphology of the photoresist. Therefore, Figure 7 shows the intensity distribution of the photoresist in the case of multiple mask holes, and the corresponding contour maps are shown in Figure 8. The mask hole spacing was 40 µm, and the mask hole size was 70 µm. The length of the photoresist was 200 µm, and the abscissa in the figure represents the horizontal position. According to the proportion calculation, it can be seen that the simulation results are correct.

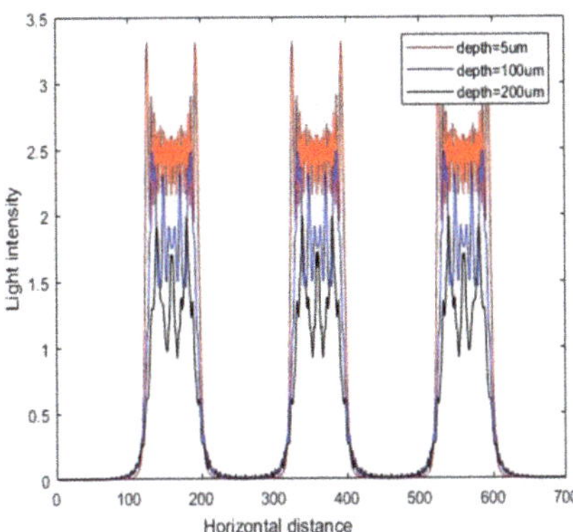

Figure 7. The intensity curve for vertical exposure with different depths.

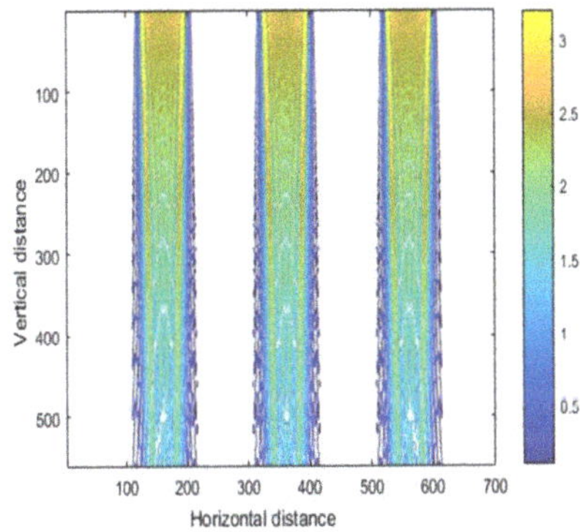

Figure 8. The contour map of intensity distribution for vertical exposure.

The above simulation results were obtained without considering the substrate reflection. Part of the light is reflected at the photoresist and substrate interfaces. The refractive index can be calculated according to the refractive index of the photoresist and the substrate, thereby enabling calculation of the distribution of light intensity after refraction. During vertical UV lithography processes of SU-8, reflection enhances light intensity at the same position as the photoresist. During inclined UV lithography processes of SU-8, the incident UV light is reflected at the SU-8/substrate interface. The reflected UV light will be transmitted into other unexposed SU-8, usually creating reflected induced structures. The light intensity distribution under different conditions at the same position is shown in Figures 9 and 10. By comparing the value of light intensity with and without reflection at the same position, we found that the light intensity considering substrate reflection was greater than that without substrate reflection, and this phenomenon was more obvious at the bottom of the photoresist. This is consistent with the previous analysis.

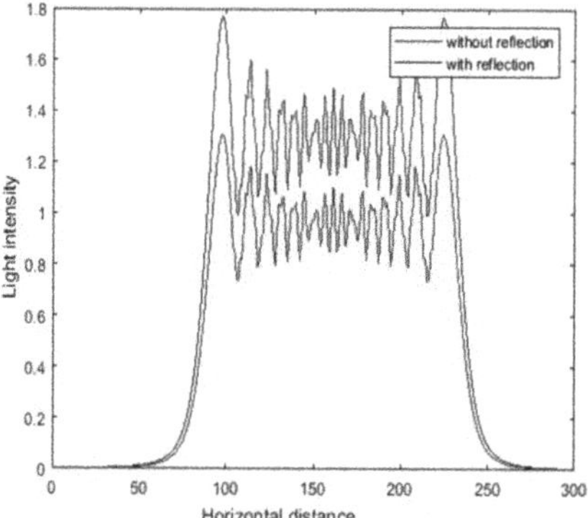

Figure 9. The intensity curve for vertical exposure at depth of 150 μm.

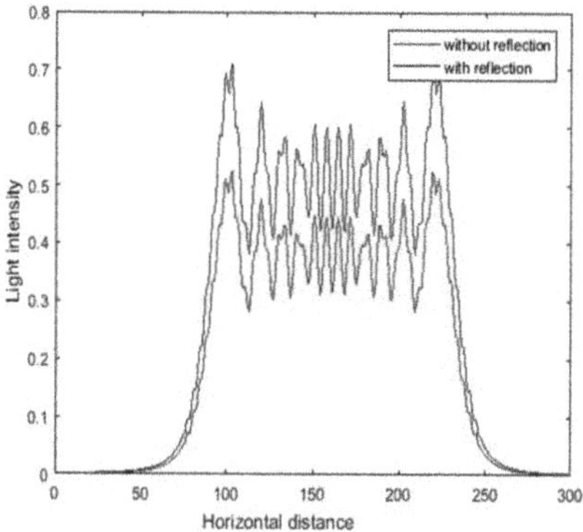

Figure 10. The intensity curve for vertical exposure at depth of 300 μm.

One of the advantages of substrate reflection is that it can ensure the photoresist bottom is fully exposed, avoiding the disadvantage of the photoresist bottom not being fully exposed due to it being too thick. It should be noted that if the substrate reflection is not well controlled, it will lead to overexposure of the photoresist bottom, which in turn will lead to deviation between the results and the design. The reflection at the photoresist/substrate interface cannot be neglected as the dose of reflected UV light is increased by increasing the exposure time to more than that of a regular exposure process. To eliminate the reflected induced patterns, we can coat a layer of antireflective film on the photoresist or substrate surface, called TARC and BARC, respectively.

Secondly, we analyzed the simulation of light intensity at oblique incidence. The incident UV light was inclined from left to right, and the incident light intensity was 2.6 mW/cm^2. Other conditions were the same as vertical incidence.

The intensity curves for inclined exposure with different air gaps are shown in Figure 11, and the corresponding contour maps are shown in Figure 12. As can be seen, as the air gap increased from 10 to 30 μm, the exposure area shifted to the right, the diffraction effect became more intense, and the edge diffraction effect also became severe.

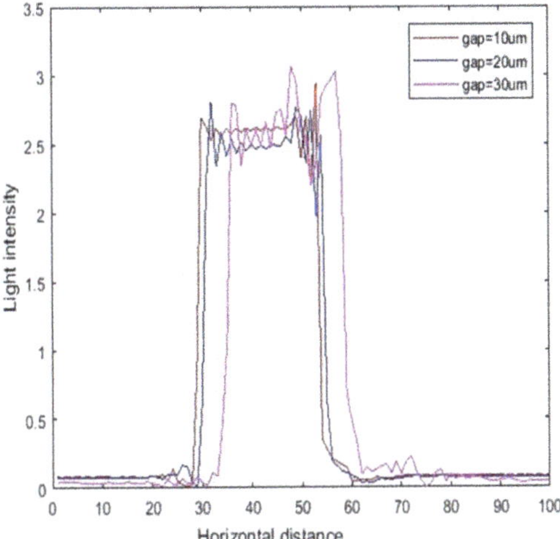

Figure 11. The intensity curve for inclined exposure with different air gaps (10, 20, and 30 μm) at 23.5° incident angle.

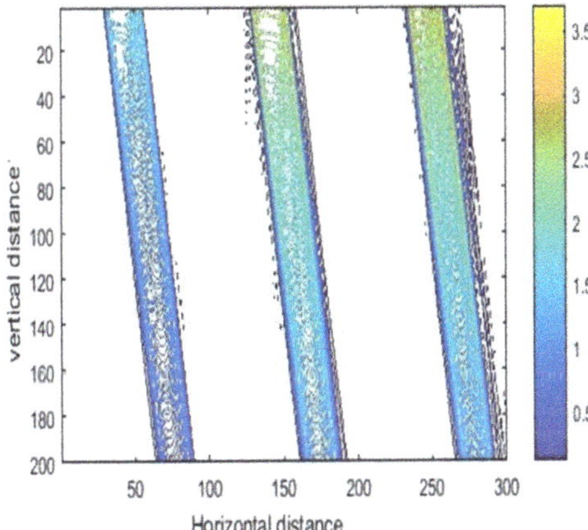

Figure 12. The contour map of intensity with different air gaps (10, 20, and 30 μm) at 23.5° incident angle.

Figure 13 illustrates the intensity curves resulting from different incident angles, and the corresponding contour maps are shown in Figure 14. We can see from the results that the exposure area shifted to the right, and the light intensity began to decrease with increasing inclined angle. Analyzing this phenomenon from a mathematical point of view, according to the expression of matrix R in Formula (8), the value of R decreased with the increase in incident angle, and therefore the final light intensity also decreased.

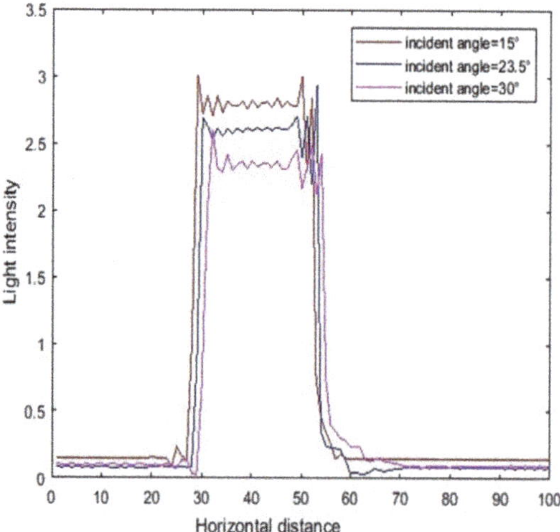

Figure 13. The intensity curve for inclined exposure with different incident angles in the X–Z cross section.

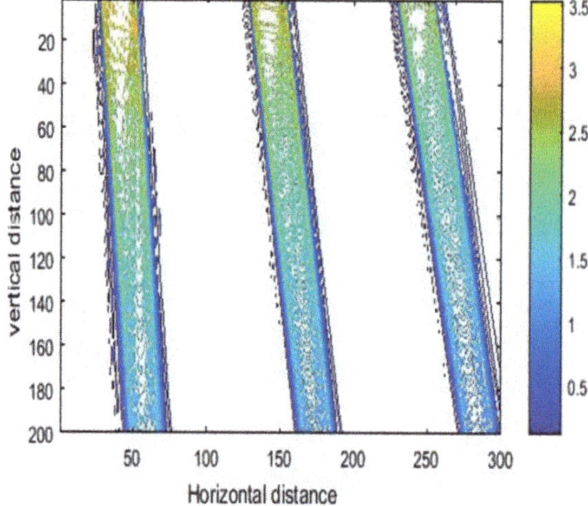

Figure 14. The contour map of intensity with different incident angles (15, 23.5, and 30°) with air gap of 10 μm.

The intensity curves for inclined exposure with different thicknesses are shown in Figure 15, and the corresponding contour maps are shown in Figure 16. It can be seen from the Figure 15 that due to diffraction of light and absorption of the photoresist, the light intensity was different at different depths. On the surface of the photoresist, the light intensity was the highest, while it was the smallest on the bottom of the photoresist. It should be noted that the intensity in SU-8 was simultaneously attenuated along the radiation direction with increasing photoresist thickness.

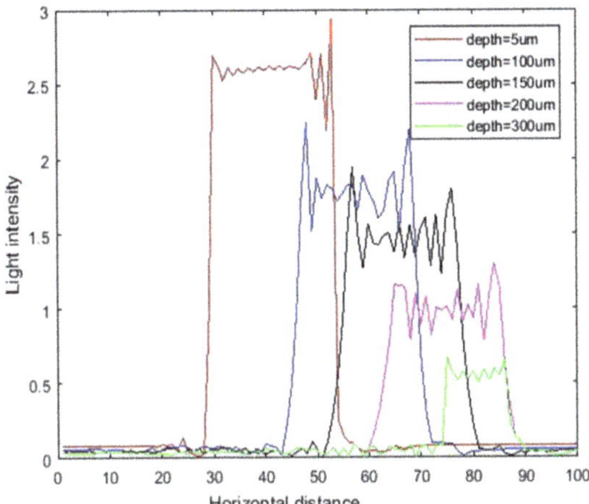

Figure 15. The intensity curve for inclined exposure with different depths at 23.5° incident angle.

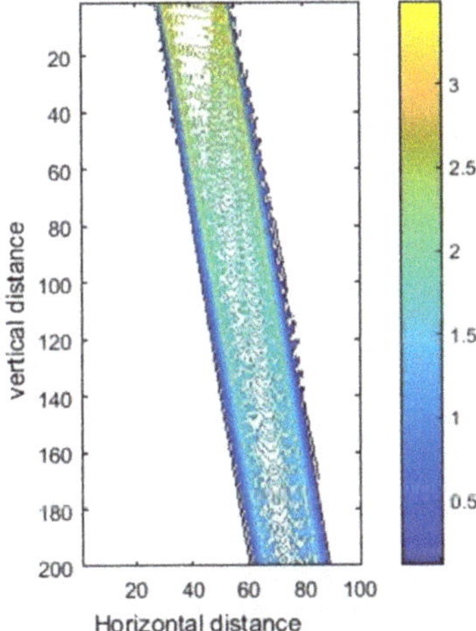

Figure 16. The contour map of intensity distribution for inclined exposure at incident angle 23.5°.

Compared with vertical incidence, the intensity curve under oblique incidence decreased from top to bottom and shifted to the right. Further simulation showed that this translation amplitude would be larger with the increase of incident angle. Finally, the computation time depended on the number of orders of the Fourier series expansion. With an increase of the expansion order, the calculation accuracy was higher and the calculation time increased accordingly. Therefore, in actual simulation, it is necessary to select the appropriate expansion order in order to consider the calculation time and calculation accuracy.

3.2. Experimental Results and Analysis

Experiments for UV lithography were conducted to validate the proposed lithography simulation model, and simulation results based on exposure intensity distribution patterns are illustrated for demonstration. In this calculation, the incident light intensity was 2.6 mW/cm^2, and its wavelength was 365 nm. The thickness of SU-8 photoresist was 300 μm, and the thickness of the air gap was 10 μm. Besides, in order to efficiently evaluate and compare the simulation and experimental results, the contour value of light intensity patterns was assumed as 0.38 in this study.

SEM photos of SU-8 structures fabricated on glass wafers are illustrated in Figure 17a, and the corresponding simulation results are shown in Figure 17b. From Figure 17a, we can see that the width of the top SU-8 was 31.37 μm, and the width of the bottom SU-8 was 32.06 μm. Therefore, after development, the width of SU-8 pillars was enlarged over the entire height of the photoresist. In particular, the variation of the bottom width was 0.76 μm. Due to the absorption of light by photoresist, for a relatively low exposure dose near the bottom of the SU-8 photoresist layer, the width at the bottom is slightly larger than that at the top. By means of threshold prediction (the contour value was assumed as 0.38) of the light intensity pattern, the top and bottom widths derived from the simulation result in Figure 17b were 33.28 and 34.7 μm, respectively. In comparison, the top width in the intensity pattern was approximately equal to that in the experimental result. Therefore, we drew the conclusion that the simulation results were in good agreement with the experimental results.

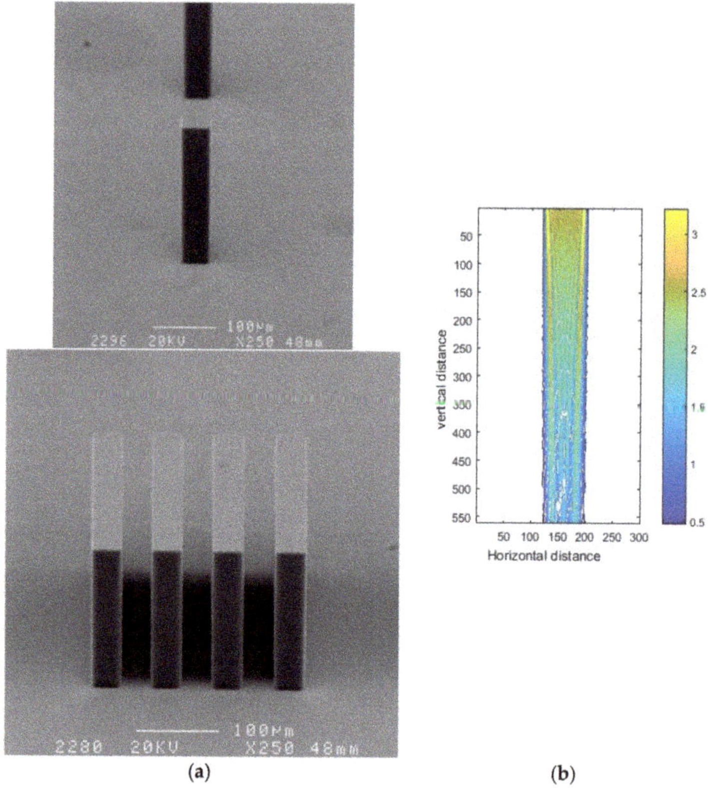

(a)   (b)

Figure 17. SEM photo of a SU-8 pillar (**a**) and simulation result (**b**) for mask hole size of 30 μm.

Figures 18 and 19 show a series of SU-8 pillars with different proportions of linewidth and spacewidth and the corresponding simulation results of exposure intensity. SEM photos of the SU-8 structures fabricated on glass wafers are illustrated in Figures 18a and 19a, and the corresponding simulation results are shown in Figures 18a and 19b. In practice, there are often multiple mask holes, and the spacing between the mask holes is an important parameter. Different mask hole spacing will affect the final development morphology of photoresist.

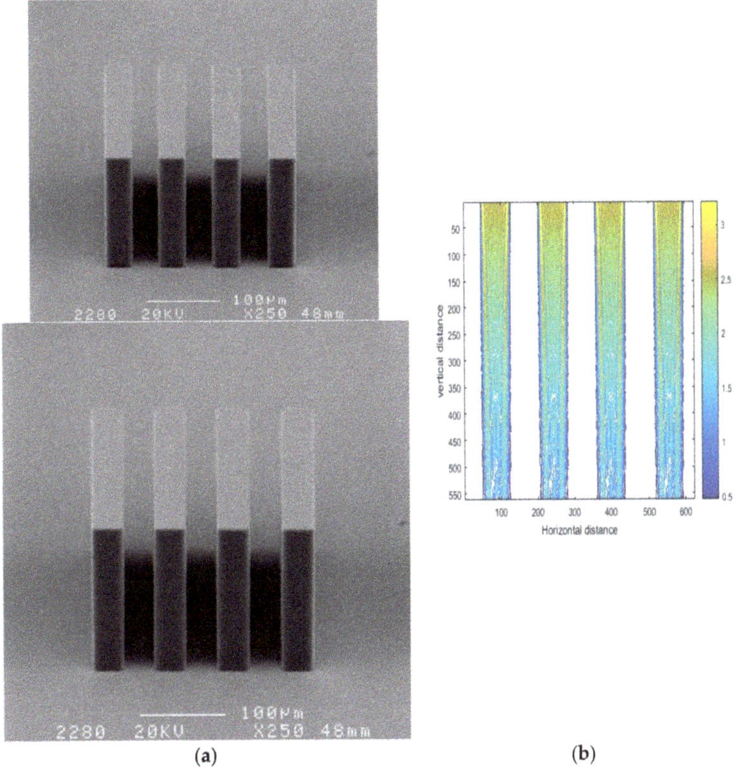

Figure 18. SEM photo of a SU-8 pillar (**a**) and simulation result (**b**) for linewidth/spacewidth of 30/30 μm.

Figure 18 shows the simulation and experimental results for linewidth/spacewidth of 30/30 μm. Through measurement and calculation, the top and bottom widths of the photoresist cylinder, shown in Figure 18a, were found to be 32.35 and 35.29 μm, respectively. Therefore, the top and bottom widths of developed SU-8 pillars were enlarged by 2.35 and 5.29 μm, respectively. Figure 19 shows the simulation and experimental results for linewidth/spacewidth of 30/50 μm. The top and bottom widths of the photoresist cylinder, shown in Figure 19a, were 31.78 and 33.08 μm, respectively. Therefore, the top and bottom widths of developed SU-8 pillars were enlarged by 1.78 and 3.08 μm, respectively. The width of the bottom of SU-8 glue is larger than that of the top. This is because the polymer is not very dense when the exposure dose at the bottom of SU-8 photoresist is relatively low. To compare the experimental and simulation results, the dimensions of simulation results were also measured according to the contour value of light intensity patterns. The top and bottom widths for linewidth/spacewidth of 30/30 μm, shown in Figure 18b, were 31.61 and 30.58 μm, respectively, while the top and bottom widths for linewidth/spacewidth of 30/50 μm, shown in Figure 19b, were 31.43 and 30.68 μm, respectively. A comparison between Figure 18a,b and Figure 19a,b showed that the simulation results exhibited a similar trend as the experimental results. However, the disparity between the experimental and simulation results in Figure 18 is notable. This was due to the decrease of mask hole spacing. The results revealed that the interaction between neighboring slits became more significant with diminution of the interval between two slits, and the diffraction effect was enhanced with a decrease in the interval between two slits, leading to the linewidth of the SU-8 structure with linewidth/spacewidth of 30/30 μm becoming a little larger than that with linewidth/spacewidth of 30/50 μm.

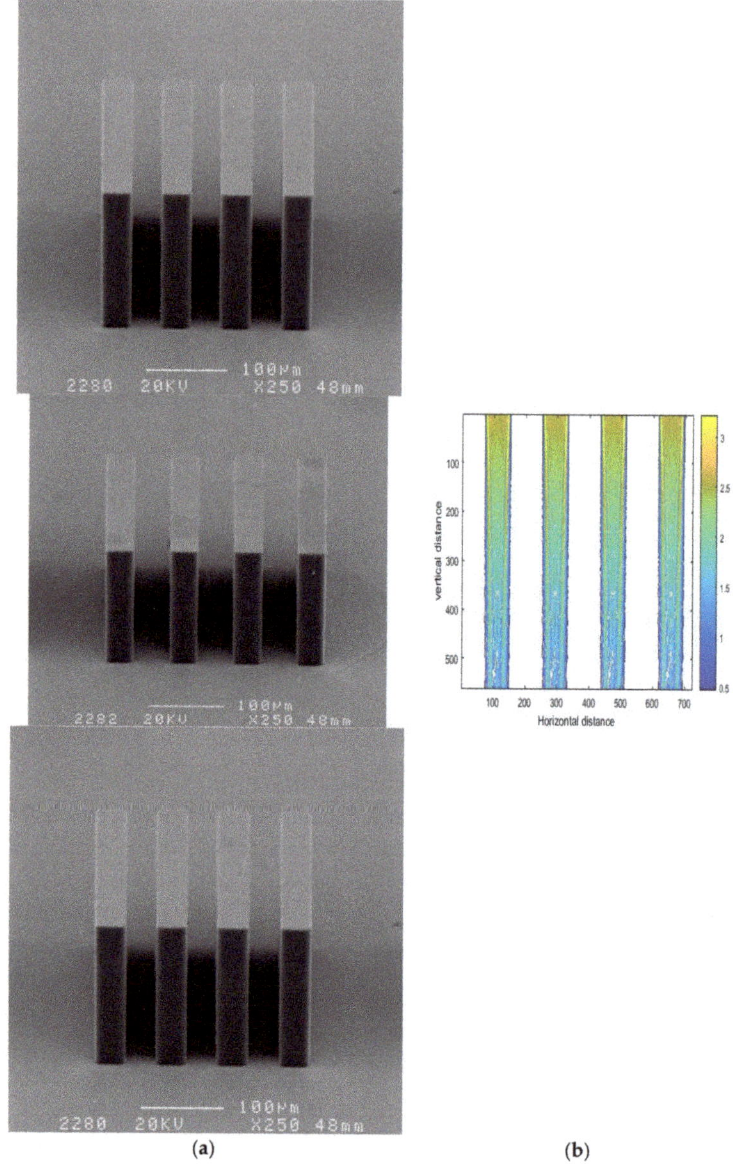

Figure 19. SEM photo of a SU-8 pillar (**a**) and simulation result (**b**) for linewidth/spacewidth of 30/50 μm.

SEM photos of the oblique SU-8 structures fabricated on glass wafers are illustrated in Figure 20a, and the corresponding simulation results are shown in Figure 20b. Figure 20a,b shows the intensity distribution of photoresist in the case of multiple mask holes with the incident angle of 30° and linewidth/spacewidth of 30/60 μm. From Figure 20a, we can see that the width of the top SU-8 was 31.75 μm, and the width of the bottom SU-8 was 33.40 μm. Therefore, the top and bottom widths of developed SU-8 pillars were enlarged by 1.75 and 3.40 μm, respectively. In particular, the variation of the bottom width was 1.75 μm. We also calculated the top and bottom widths in the simulation results. In comparison, the top width in the intensity pattern was approximately equal to that in the experimental result. Therefore we drew the conclusion that the simulation results were in good agreement with the experimental results. The measured structural inclined angles of developed SU-8 photoresists were 73.76° for the incident angle of 30°. Correlating Formula (12), the theoretical structural inclined angle α of developed structures was 72.58° for the incident angle of 30°, which was approximately the same as the measured value and thus demonstrated good agreement with the experimental results.

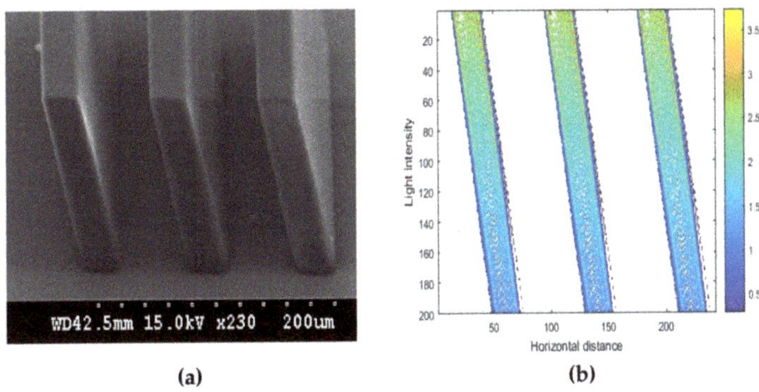

Figure 20. SEM photo of an oblique SU-8 pillar (**a**) and simulation result (**b**) for incident angle of 30° and linewidth/spacewidth of 30/60 μm.

4. Conclusions

In order to simulate UV lithography and predict the profiles of SU-8 structures, a lithography simulation model based on the waveguide method and rigorous electromagnetic field theory is presented in this paper. According to this model, vertical and inclined UV exposure of SU-8 thick photoresist was successfully simulated, and the parameters significantly influencing light intensity distribution were studied. The presented model using the WG method works faster than other electromagnetic field methods and produces precise results. A series of experiments were designed to investigate the performance of the simulation model. By comparing simulation and experiments results, the method was successfully verified. This model is ideally suited for 2D lithography simulations for complex cases.

Author Contributions: Z.-C.G. and Z.-F.Z. developed the model and designed the experiments; Z.-C.G. performed the experiments; H.D. analyzed the data; Q.-A.H. proposed the idea for the model and made suggestions for improvement. Z.-C.G. and Z.-F.Z. edited this manuscript. All authors have read and agreed to the published version of the manuscript.

Funding: This study was supported by the National Key R&D Program of China under grant number 2018YFB2002600.

References

1. Cui, Z.; Jenkins, D.W.; Schneider, A.; McBride, G. Profile control of SU-8 photoresist using different radiation sources. In *MEMS Design, Fabrication, Characterization, and Packaging*; International Society for Optics and Photonics (SPIE): Bellingham, WA, USA, 2001; Volume 4407, pp. 119–126.

2. Arscott, S. SU-8 as a material for lab-on-a-chip-based mass spectrometry. *Lab Chip* **2014**, *14*, 3668–3689. [CrossRef] [PubMed]

3.	Lorenz, H.; Despont, M.; Fahrni, N.; Brugger, J.; Vettiger, P.; Renaud, P. High-aspect-ratio, ultrathick, negative-tone near-UV photoresist and its applications for MEMS. *Sens. Actuators A Phys.* **1998**, *64*, 33–39. [CrossRef]

4.	Gray, B.L.; Lordanov, V.P.; Sarro, P.M.; Bossche, A. SU-8 Structures for Integrated High-Speed Screening. In *Micro Total Analysis Systems*; Baba, Y., Shoji, S., van den Berg, A., Eds.; Springer: Dordrecht, The Netherlands, 2002; pp. 464–466. [CrossRef]

5.	Jackman, R.J.; Floyd, T.M.; Ghodssi, R.; Schmidt, M.; Jensen, K.F. Microfluidic systems with on-line UV detection fabricated in photodefinable epoxy. *J. Micromech. Microeng.* **2001**, *11*, 263–269. [CrossRef]

6.	Lorenz, H.; Despont, M.; Fahrni, N.; LaBianca, N.; Renaud, P.; Vettiger, P. A low cost negative resist for MEMS. *J. Micromech. Microeng.* **1997**, *7*, 121–124. [CrossRef]

7.	Neureuther, A.W. Understanding Lithography Technology Issues through Simulation. *Jpn. J. Appl. Phys.* **1993**, *32*. [CrossRef]

8.	Levinson, H.J. *Principles of Lithography*; SPIE Press: Bellingham, WA, USA, 2005.

9.	Mack, C.A. PROLITH: A Comprehensive Optical Lithography Model. In Proceedings of the 1985 Microlithography Conferences, Santa Clara, CA, USA, 11–14 March 1985; Volume 538, pp. 207–220.

10.	Born, M.; Wolf, E. *Principles of Optics: Electromagnetic Theory of Propagation, Interference and Diffraction of Light*; Cambridge University Press: Cambridge, UK, 1999.

11.	Rudolph, O.H.; Evanschitzky, P.; Erdmann, A.; Bär, E.; Lorenz, J.K. Rigorous electromagnetic field simulation of the impact of photomask line-edge and line-width roughness on lithographic processes. *J. Micro/Nanolithog. MEMS MOEMS* **2012**, *11*, 013004-1. [CrossRef]

12.	Evanschitzky, P.; Erdmann, A. Fast near field simulation of optical and EUV masks using the waveguide method. In Proceedings of the 23rd European Mask and Lithography Conference, Grenoble, France, 22–26 January 2007.

13.	Erdmann, A.; Evanschitzky, P.; Citarella, G.; Fühner, T.; De Bisschop, P. Rigorous mask modeling using waveguide and FDTD methods: An assessment for typical hyper-NA imaging problems. In *Photomask and Next-Generation Lithography Mask Technology XIII*; International Society for Optics and Photonics (SPIE): Bellingham, WA, USA, 2006; Volume 6283, p. 628319. [CrossRef]

14.	Nyyssonen, D. Theory of optical edge detection and imaging of thick layers. *J. Opt. Soc. Am.* **1982**, *72*, 1425–1436. [CrossRef]

15.	Yuan, C.M. Calculation of one-dimensional lithographic aerial images using the vector theory. *IEEE Trans. Electron Devices* **1993**, *40*, 1604–1613. [CrossRef]

16.	Evanschitzky, P.; Erdmann, A. Three dimensional EUV simulations: A new mask near field and imaging simulation system. *Photomask Technol.* **2005**, *5992*. [CrossRef]

17.	Zhu, Z.; Lucas, K.; Cobb, J.L.; Hector, S.D.; Strojwas, A.J. Rigorous EUV mask simulator using 2D and 3D waveguide methods. In *Emerging Lithographic Technologies VII*; International Society for Optics and Photonics (SPIE): Bellingham, WA, USA, 2003; Volume 5037, pp. 494–503.

18.	Yuan, C.-M. Efficient light scattering modeling for alignment, metrology, and resist exposure in photolithography. *IEEE Trans. Electron Devices* **1992**, *39*, 1588–1598. [CrossRef]

19.	Lucas, K.D.; Tanabe, H.; Strojwas, A.J. Efficient and rigorous three-dimensional model for optical lithography simulation. *J. Opt. Soc. Am. A* **1996**, *13*, 2187–2199. [CrossRef]

20.	Schellenberg, F.M.; Adam, K.; Matteo, J.; Hesselink, L. Electromagnetic phenomena in advanced photomasks. *J. Vac. Sci. Technol. B Microelectron. Nanometer Struct.* **2005**, *23*, 3106. [CrossRef]

21.	Shao, F.; Evanschitzky, P.; Reibold, D.; Erdmann, A. Fast rigorous simulation of mask diffraction using the waveguide method with parallelized decomposition technique. In Proceedings of the 24th European Mask and Lithography Conference, Dresden, Germany, 21–24 January 2008.

22.	Tanabe, H. Modeling of optical images in resists by vector potentials. In *Optical/Laser Microlithography V*; International Society for Optics and Photonics (SPIE): Bellingham, WA, USA, 1992; Volume 1674, pp. 637–649.

23.	Yuan, C.-M.; Strojwas, A.J. Modeling optical microscope images of integrated-circuit structures. *J. Opt. Soc. Am. A* **1991**, *8*, 778. [CrossRef]

24.	Gaudet, M.; Arscott, S. A user-friendly guide to the optimum ultraviolet photolithographic exposure and greyscale dose of SU-8 photoresist on common MEMS, microsystems, and microelectronics coatings and materials. *Anal. Methods* **2017**, *9*, 2495–2504. [CrossRef]

25. Wang, F.; Zhou, Z.; Weihua, L.; Qing'an, H. Rigorous electromagnetic field model for optical lithography simulation. *High Power Laser Part. Beams* **2015**, *27*, 024106. [CrossRef]

26. Wojcik, G.L.; Vaughan, D.K.; Mould, J.J.; Leon, F.A.; Qian, Q.-D.; Lutz, M.A. Laser alignment modeling using rigorous numerical simulations. In *Optical/Laser Microlithography IV*; International Society for Optics and Photonics (SPIE): Bellingham, WA, USA, 1991; Volume 1463, pp. 292–303.

27. Gaudet, M.; Camart, J.-C.; Buchaillot, L.; Arscott, S. Variation of absorption coefficient and determination of critical dose of SU-8 at 365 nm. *Appl. Phys. Lett.* **2006**, *88*, 24107. [CrossRef]

Publisher's Note: MDPI stays neutral with regard to jurisdictional claims in published maps and institutional affiliations.

 *micromachines*

Article

The Micro Topology and Statistical Analysis of the Forces of Walking and Failure of an ITAP in a Femur

Euan Langford [1], Christian Andrew Griffiths [2],*, Andrew Rees [2] and Josh Bird [3]

1. The Defence Science and Technology Laboratory (DSTL), Fareham PO17 6AD, UK; elangford@dstl.gov.uk
2. College of Engineering, Swansea University, Swansea SA1 8EN, UK; Andrew.Rees@Swansea.ac.uk
3. Department of Mechanical Engineering, University of Bath, Bath BA2 7AY, UK; jb3315@bath.ac.uk
* Correspondence: c.a.griffiths@swansea.ac.uk

Abstract: This paper studies the forces acting upon the Intraosseous Transcutaneous Amputation Prosthesis, ITAP, that has been designed for use in a quarter amputated femur. To design in a failure feature, utilising a safety notch, which would stop excessive stress, σ, permeating the bone causing damage to the user. To achieve this, the topology of the ITAP was studied using MATLAB and ANSYS models with a wide range of component volumes. The topology analysis identified critical materials and local maximum stresses when modelling the applied loads. This together with additive layer manufacture allows for bespoke prosthetics that can improve patient outcomes. Further research is needed to design a fully functional, failure feature that is operational when extreme loads are applied from any direction. Physical testing is needed for validation of this study. Further research is also recommended on the design so that the σ within the ITAP is less than the yield stress, σ_s, of bone when other loads are applied from running and other activities.

Keywords: prostheses; ITAP; micro topology; ANSYS; MATLAB; additive manufacture

 check for **updates**

Citation: Langford, E.; Griffiths, C.A.; Rees, A.; Bird, J. The Micro Topology and Statistical Analysis of the Forces of Walking and Failure of an ITAP in a Femur. *Micromachines* **2021**, *12*, 298. https://doi.org/10.3390/mi12030298

Academic Editors: Davide Masato and Giovanni Lucchetta

Received: 17 February 2021
Accepted: 11 March 2021
Published: 12 March 2021

Publisher's Note: MDPI stays neutral with regard to jurisdictional claims in published maps and institutional affiliations.

1. Introduction

The failure features of the femur are unique to the human skeleton because the bone is designed to transfer σ throughout the structure when compressive or tensile loads are applied. The σ is also dispersed to the surrounding muscle, cartilage and connected bone in the skeleton. This allows the femur to handle large σ when the user is in motion such as running and jumping. However, the femur is not indestructible as the bone structure is susceptible to fracture from extreme shear loads. It has been theorised by that a sheer force of 4000 N applied perpendicular to the bone would cause a fracture [1]. The properties of the individual's bone structure are also key factors for the propagation of fracturing as the physiology is unique to the individual. This means that some individuals and ages are more susceptible to fractures particularly when bone decreases after the age of 30 [2]. The femur can also fail under different extreme loads from, for example, collision or devices such as landmines, improvised explosive devices, IEDs, and other explosives as identified by Mckay et al. [3]. The direction and position of the applied σ is difficult to define, because the load could be applied from any direction at any time.

The intraosseous transcutaneous amputation prosthesis, ITAP, is a direct skeleton attachment that is used by amputees [4]. This is used rather than a standard socket interface that causes swelling, discomfort and potential infection for the user. The socket interface often prohibits the life span of the use of the prosthetic [5]. As a result, the development of the ITAP and similar direct skeleton attachments have been designed to mitigate these problems, to improve the quality of life for the user [4]. ITAPs were initially designed from studying animals with protruding bone such as deer and rhinos [6]. The mechanics and biomechanics of these animals' bone development have been analysed due to the lack of infection of the tissue surrounding the protruding bone which is similar to an embedded prosthetic. This is vital, as an embedded prosthetic can be susceptible to infection and damage of the local tissue and if a prosthetic is not properly manufactured with biomedical procedures with appropriate postproduction methods [4]. Consequently, the ITAP is constructed in a manner that replicates the antlers of deer. An open insertion to the bone with reduced bacterial colonization infection was realised using a diamond-like carbon (DLC) ITAP [6].

The biomechanics of the ITAP and the human body need to be thought of during design, simulation and production as the material must not be harmful to the patient's body. As such, the main material used is that of Titanium (Ti). This is because Ti is biocompatible with the human body, with a high processability through Computer-aided design (CAD) and additive manufacturing (AM) processes. Ti is also strong enough to handle the loads associated with walking and other movements [7,8]. However, it can deteriorate in to smaller fragments in the body over time due to movement. This is most commonly attributed to hip and knee implants because there is constant friction between the implant and the joint [9]. Additional alternative materials can be used for prosthetics such as Polyetheretherketone (Peek). This can be processed by AM methods and inserted into the bone to help repair and supplement the bone after injury. This is commonly used for American football players for bone repair therapy as Peek is resistant to infection [10] However, Peek is not suitable for use in a load bearing ITAP as there is poor bonding between the bone and the material. Therefore, it could easily become possible for the ITAP to be partially or fully dislodged from the bone, causing internal damage to the user.

The ITAP can be used for both cosmetic and load bearing prosthetics; the former can range from prosthetic eyes, fingers, noses, ears amongst others [11], while the latter can be used for prosthetic limbs with loadbearing forces acting upon extremities. This includes, but is not limited to, both high and low leg amputations, either above or below the knee However, the work of Newcombe et al., identified that an ITAP implant was not advisable for an amputation greater than one quarter of the original length of the bone. This was due to the residual σ in the femur housing, known as the anchor, being too great so damage to the cortical bone could occur [12]. Further suggested amendments to the ITAP have been developed to reduces the σ with in both the bone and the ITAP. This includes a safety notch that is positioned outside the user's congealed tissue. The proposed modification could prevent an extreme load from damaging the user as the ITAP would fallure at the safety notch [13].

Other direct skeletal prosthetics are available for use such as Osseointegrated Prostheses for the Rehabilitation of Amputees, OPRA. This system involves screwing into the bone cavity which increases the residual σ all along the anchor. This system works because it increases the user's hip flexibility and extension in comparison to traditional socket interfaces. However, over time it was recorded that pain was detected due to the build-up of residual σ [14]. The ITAP is not only suited to use in humans but it has had clinical trial in canines. Noel Fitzpatrick et al. studied the procedure for installation and use in four separate canines. The study identified that all subjects' quality of life improved due to the prosthetic attachment. However, canines 1–3 were euthanized due to metastatic disease spreading throughout the body; it was unclear if this was related to the ITAP [15]. As the ITAP can be fitted for different purposes and species, bespoke parts can be produced. This is extremely useful as the properties of bone vary as well as the length of the remaining bone after amputation, see Figures 1 and 2. Bone shape and modelling can be recorded by using Computed Tomography (CT) scanning. This creates a need for design iterations to model and build an ITAP that suits the individual's lifestyle. In this process it is essential to design a build that does not over engineer the forces acting upon the leg. Failure to work in this design constraint could cause near fatal damage to the user resulting in further amputation or loss of life. This is due to the failure of the anchor supporting the ITAP and the potential rupturing of arteries and veins surrounding the localized area of the prosthetic. The work of Sullivan et al. concluded that the use of direct skeleton attachments for prosthetics increases the user's quality of life and the function of the prosthetic limb. Comfort also increased due to the lack of sores developing on the stump. This is because there is no contact between the stump of the limb and the prosthetic sock fitted over the top. However, the reports concluded that aspects of the development of the ITAP needed further advancement because the "trans-femoral amputee" case needs extensive review before it becomes a common clinical procedure for prosthetic rehabilitation [16]. It has been identified by Bird et al., that the initial area of failure and high σ concentration of the ITAP is located at the root of the ITAP [13]. The results of the study identified that a single safety notch was not enough for the ITAP to be suitable for use. This is because the strength of the Ti used for the ITAP is greater than the bone, so the bone would fail before the ITAP, resulting in damage to the user. Thus, it is identified that a comparable σ between bone and the ITAP at the point of insertion is needed for successful development of the product. This is categorized as designing in failure for success.

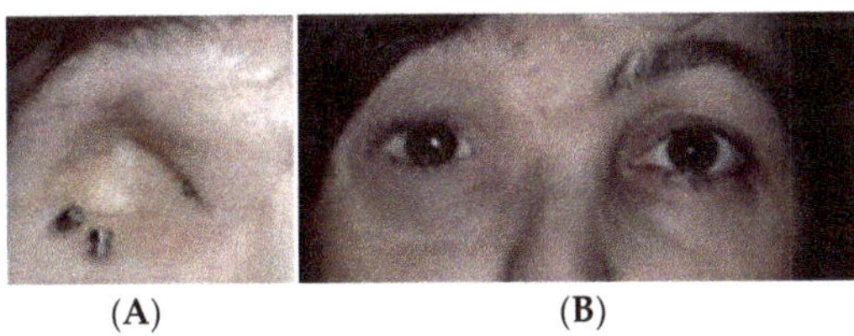

(A)   **(B)**

Figure 1. (**A**) ITAPs embedded into the skull for a prosthetic eye, (**B**) Prosthetic eye mounted on the ITAPs into the skull [11].

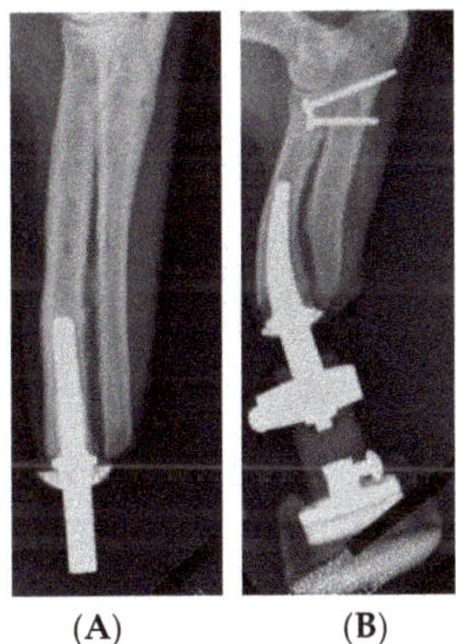

(A)   **(B)**

Figure 2. (**A**) Canine prosthetic ITAP embedded into its front left leg. (**B**) A Canine bespoke ITAP embedded into its front right leg [15].

The benefit of using AM methods of production is that the topology of the ITAP can be adapted to meet the users bespoke requirement. The manufacturing process allows for modification of the internal structure, of the ITAP and each can be printed to have the same properties and micro topology as the bone being amputated, as seen via CT scanning. The first aim of this study is to identify the optimum topology of the ITAP, locating critical material that is integral to the structure of the design when simulated to both walking and extreme loads. The secondary aim is to optimise the design of the ITAP topology so that a controlled failure can be achieved when extreme loads are applied. Thus, producing an ITAP that prevents a further damage to bone when subjected to excessive loads. For both aims the simulation will consider the design of a quarter amputation of a femur with an ITAP embedded into the bone (Figure 3). Both walking and extreme σ studies are considered because the operational σ of the design needs to be evaluated in order to design a safe prosthetic that can be embedded into the bone. To ensure reliable results a multiple modelling software approach is used to generate the topologies. The paper is organised as follows: the next section uses simulation methods to identify the maximum stress, σ_{Max} on the standard ITAP due to walking loads. This is followed by topology optimisation of the ITAP model using ANSYS and MATLAB. Then, Section 4. considers the influence of the topology of the ITAP design when under extreme loads. Finally, conclusions are presented on the optimisation of ITAP designs.

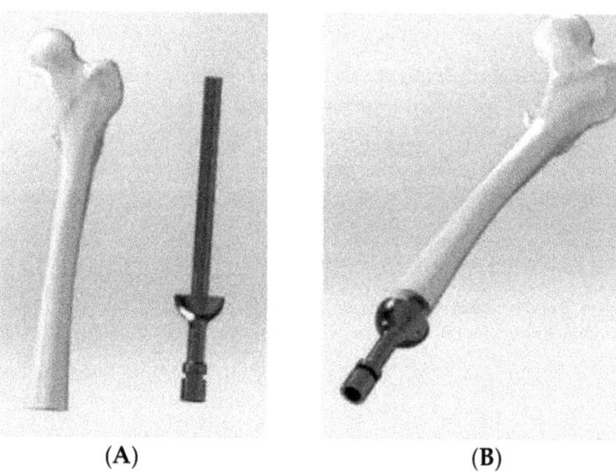

Figure 3. (**A**) Quarter Amputated Femur with accompanying ITAP CAD models, (**B**) Femur and ITAP assembly [17].

2. Experimental Methodology

The purpose of the research is to establish the strength of the ITAP, and to identify the behaviour of a safety mechanism to prevent excessive forces transferring to bone. The focus will be on performing Simulations of σ_{Max} on the ITAP due to walking loads. In order to perform this the following section will identify initial design, boundary conditions and mesh used. In 2.2 the Simulated results of the ITAP standard design will be presented and then used as a benchmark for the ITAP designs with topology optimisation in mboxsectsect:sec3-micromachines-1132465.

2.1. The Initial Design, Boundary Conditions and Mesh

The A quarter amputation of the femur requires an ITAP with an embedded depth of 160 mm, with a diameter of 14 mm [12]. The femur provided in this study was obtained by CT scanning of a deceased male aged 44 weighing at 85 kg and at a height of 185 cm [17]. Based on this femur the standard ITAP design can be seen in Figure 3 The ITAP has been segmented into 4 section as to identify the areas of the recommended change of the topology, when simulated with the loads associated with walking and bone failure. This has been done as it is imperative that the ITAP bonds to the bone and allows the muscle to congeal around the protruding end. As such the changing topology must be correctly identified and implemented, in different section to reduce the risk of failure of insulation. As the designed in failure mechanic must failure externally of the body. Thus, the segments utilized are Section A, B, C and D theses represent the region of the ITAP: embedded in to the femur that must have a constant surface contact with bone; located at the end of the users stump where the collecting skin, fat and muscle fusses; were the desired failure mechanic is needed to be implemented for a safe failure of the ITAP; were the prosthetic leg is attached to the user, respectively. For the simulation of the design the following assumptions are mad.

- The ITAP and femur are both cylindrical, where the femur has an outer diameter of 28 mm and is extruded for 311.5 mm. Whilst there is often a curve in the femur, the tensor faciae latae muscle can reduce the stress acting along the bone and therefore reduce any pronounced stress concentrations acting along the femur [18].
- At one end of the femur there is a 14 mm diameter hole, centred in the middle, cut to a depth of 160 mm.
- The material properties of the femur have been assumed to be cortical bone, which has a σ_s of 110 and 120 MPa for compressional and tensional forces. The mechanical strength of a femur depends on numerous factors, such as age, porosity and mineral content. However, Marco et al. [19] and Reilly and Burstein [20] found average elastic and shear moduli of human femur specimens which can be used as a preliminary step towards modelling the bone housing ITAP implant. A summary of these mechanical properties is presented by Bird et al. [13].
- The material used for the ITAP has been set as Ti from the ANSYS's database, this has a σ_s of 930 MPa and an ultimate tensile stress, UTS, of 1070 MPa. The material selection for the computational modelling is in line with the Ti material used for the ITAP, thus the modelling is a comparable study.
- As the focus of the simulation is on the ITAP, the femur is fixed in place to the pelvis.
- The femur will not be simulated and the free body diagram (FBD) will only consist of the ITAP and the resulting σ is analysed for the impact on the bone anchor.

The mesh was developed by conducting a mesh sensitivity study. Ten simulations were carried out with different mesh refinements to determine the following σ results:

- σ_{Max} across the entire geometry
- σ_{Max} recorded in the core of the ITAP using a set pathway in the centre of the cylinder
- σ_{Max} recorded in the core of Section B only using a pathway control.

From the study the optimum controls for the simulation where the following;

- A face sizing on both ends of each section of the ITAP, measured at 2 mm.
- A face sizing on the cylindrical surface of each section of the ITAP, measured at 2 mm.
- An edge sizing on both edges of the ITAP sections, divided by 25.

This process removed errors and many delocalisations within the mesh to ensure each section is uniform and in alignment with each other. However, when reviewing the full geometry and specific section minor desolations where identified. This is due to the simulated model being constructed from separate sections, forming a full geometry, rather than one single section. This purpose of this was to allow an in-depth review of each section of the ITAP and change the failure mechanic within section C of the ITAP.

The simulation boundary locations have been set at the femur joint with the pelvis. Frontal (F_x) Lateral (F_y) and Axial forces (F_z) are acting on the ITAP in the region where the prosthetic would be applied as this is the same area as the patellar surface where the knee would be attached [19], as seen in Figure 4.

The work of Georg et al. identified that the forces acting on the femur are dependent on the location of interest within the bone, the body weight (BW) of the user and finally the position of the motion of walking [21]. As such, the F_x, F_y and F_z maximum forces are calculated as 958.9 N, -833.8 N and 3168.6 N, respectively.

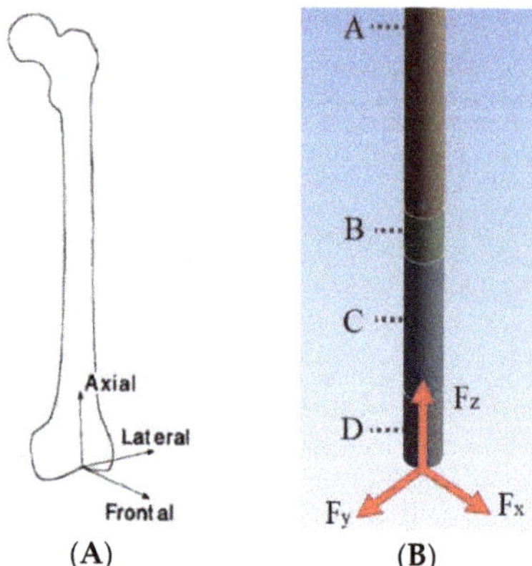

Figure 4. (**A**) Forces acting upon the femur along the bone's structure [21], (**B**) FBD of the ITAP, built into Sections A, B, C and D.

2.2. Simulated Results of the ITAP Standard Design

The σ_{Max} recorded when the ITAP is subjected to walking loads is shown in Table 1. It can be identified that as the diameter of the safety notch decreases from 14 mm to 5 mm the σ_{Max} increases. It can also be identified that the location of the σ_{Max} within the ITAP changes its position as the size of the notch diameter decreases, moving closer to the notch from the point of insertion. Discrepancies arise when analysing the full geometry's core and only Section C's core, this is partly due to minor delocalisation in the mesh within the model, further improvements have been utilised for additional readings. However, it can be identified that the recorded σ in Section C's core for all simulated models does not exceed the σ_s of Ti. Further conclusions can be drawn upon from this preliminary study, these being:

- The basic ITAP design has regions of σ above 930 MPa, with a σ_{max} of 1175.6 MPa at the A-B joint interface. This is greater than the σ_s of the Ti, and could result in a section of the ITAP permanently deforming and making it unsuitable for use.
- For all simulated geometries the section A σ is greater than both the compressive (110 MPa) and tensile (120 MPa) strength. This only occurs at the very start of the embedded part of the ITAP.

The results indicate that the ITAP design needs further optimisation to reduce the σ at the joint of bone and ITAP to less than the σ_s of bone, as well reducing the σ across the whole design so that the σ_{Max} is less than the σ_s of Ti. Topology analyses shall be utilised as the ITAP must fail within the core of Section C when the σ generated by external loads are greater than the limits of the system, thereby avoiding irreparable damage to the user. However, the core and ITAP must not fail under standard forces generated in day to day life such as walking, as seen in a 5 mm safety notch were the recorded σ is at 3313.8 MPa.

Table 1. Simulated σ_{Max} of the ITAP designs.

ITAP Designs	Standard	13 mm Safety Notch	9 mm Safety Notch	5 mm Safety Notch
σ_{Max} [MPa] In the model	1175.6	939.5	939.5	3313.8
σ_{Max} [MPa] Core of the ITAP	79.5	74.5	146.1	1003.6
Vertical height from applied loads, z domain [mm]	104.5	104.5	55.1	55.0
σ_{Max} [MPa] in Section A	570.3	735.6	735.6	735.6
σ_{Max} [MPa] Core of Section A	67.1	72.1	72.1	72.1
Vertical height from applied loads, z domain [mm]	107.4	107.4	107.4	107.4
σ_{Max} [MPa] in Section C	422.6	474.5	861.6	3313.8
σ_{Max} [MPa] Core of Section C	29.5	30.8	150.3	895.4
Vertical height from applied loads, z domain [mm]	64.7	49.9	53.3	59.0

3. Topology Optimisation of for the ITAP Model

In the following section the approach to using topology modifications to the ITAP will be show. Firstly, the ANSYS model will be described followed by the 2D and 3D approaches using MATLAB functions.

3.1. ANSYS Model

With a material reduction (50%) the design of the standard ITAP and notched designs have been simulated in ANSYS. In total 20 different simulations where conducted utilising the topology reduction calculations in order to identify key structural information for the IP when underload. The model mass is reduced in areas that are not critical to the mechanical structure, as seen in Figures 5–7. The same forces identified in Section 2.2 are used and the fixed supports used are as follows.

- The connecting surfaces of the ITAP and bone.
- The connection of the top surface of the ITAP and bone.
- The top of Section C joined to Section B of the ITAP (Figure 8D).

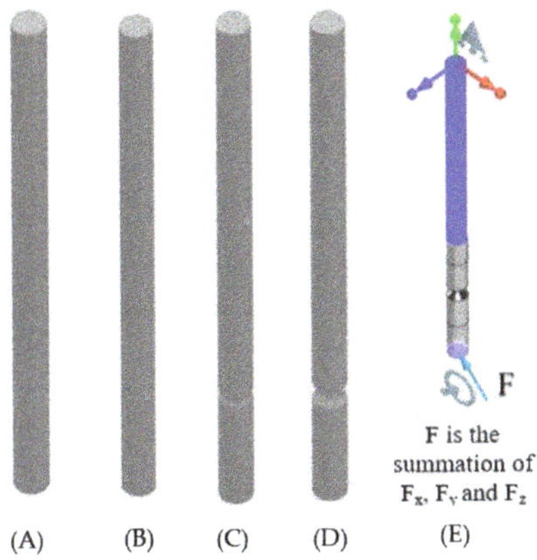

Figure 5. Fixed contact between the bone-anchor and ITAP, (**A–D**) Standard ITAP to 5 mm safety notch topology reduction models results, (**E**) FBD of the applied walking loads.

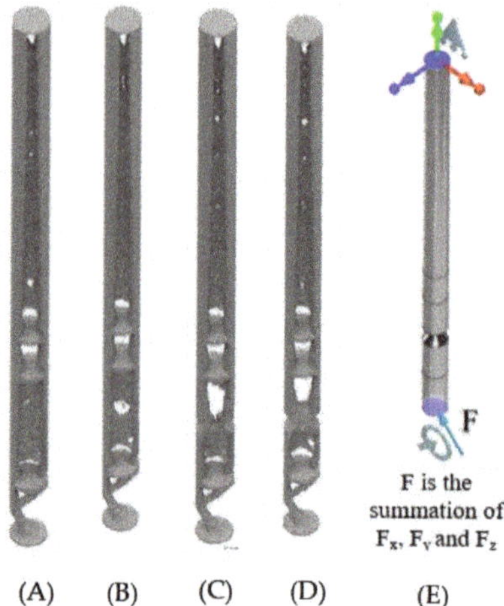

Figure 6. Fixed contact for the head of the bone-anchor to ITAP, (**A–D**) Standard ITAP to 5 mm safety notch topology reduction models results, (**E**) FBD of the applied walking loads.

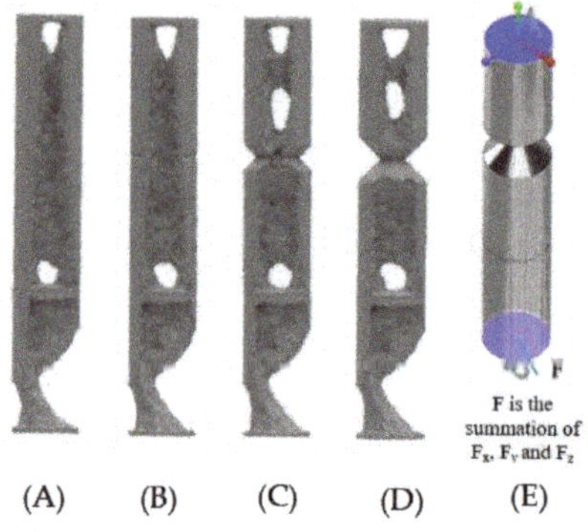

Figure 7. Fixed contact for the head of Sections B and C of the ITAP, (**A–D**) Standard ITAP to 5 mm safety notch topology reduction models results, (**E**) FBD of the applied walking loads.

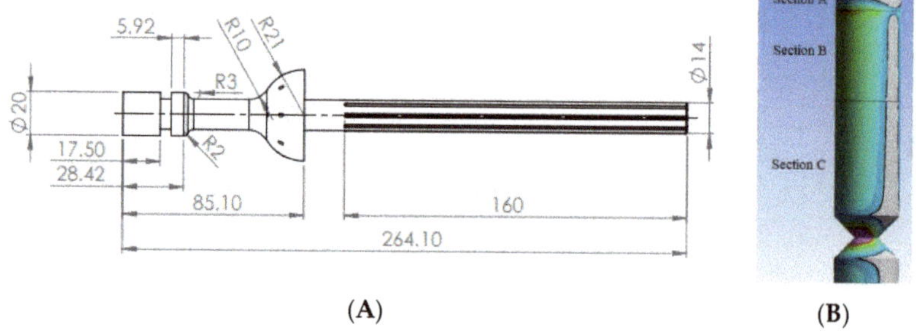

Figure 8. (**A**) Initial ITAP designed, (**B**) σ distribution in the 5 mm safety notched ITAP.

3.2. 2D Simulations Using MATLAB Functions

In this section a 2D infill for the entirety of the interior of the ITAP geometry is developed. The coding for designing an optimised interior where material can theoretically be removed originated as a 2D MATLAB problem written by Otomori et al. [22]. This has since been utilised to identify if the ANSYS topology studies identify the correct locations to remove non-critical material from the cylindrical cross-section of the ITAP's external Sections C and D. The volume of material remaining is denoted by V_{max}, and is the only variable in these simulations. Different iterations were completed looking at Vmax (50–95% in intervals of 5%), in total 10 different simulations were conducted to complete this section of research. Using MATLAB code for a level set-based topology optimization the assumptions made in the model are as follows;

- Function control levelset88 (Nelx, Nely, Vmax, Tau) are used
- The simulation considers the cross-section of the ITAP, external sections away from the bone and fusing muscle.
- The length X and Y are the length and diameter of the external region of the ITAP, Sections C and D, (represented by Nelx and Nely, respectively).
- The position of the applied load, interconnectivity and complexity is denoted by Tau and represents the regularization parameters which are set to the recommended value of 2×10^{-4} [21].
- The magnitude of the forces is not considered.
- The Young's modulus for the material is set to 1, whereas the Young's modulus of the voids has been set to 0.
- Poisson's Ratio is set at 0.3.

The resulting design structures (Figure 9) show that as the maximum volume of the model increases, the size of the voids forming in the ITAP decrease. This is most prominent in the voids forming at the boundary wall and the voids forming where sections C and D meet. This study has identified where the areas of the topology of the ITAP should be changed depending on the volume of the material used in additive manufactured production processes for complex internal structures. The topology simulation using MATLABs' applied load only consisted of one shear force, rather than the three forces that are applied to the ITAP in motion. Furthermore, the force cannot be set to the applied loads identified in Section 2.1. As such, the results from this study can only be used as a guideline.

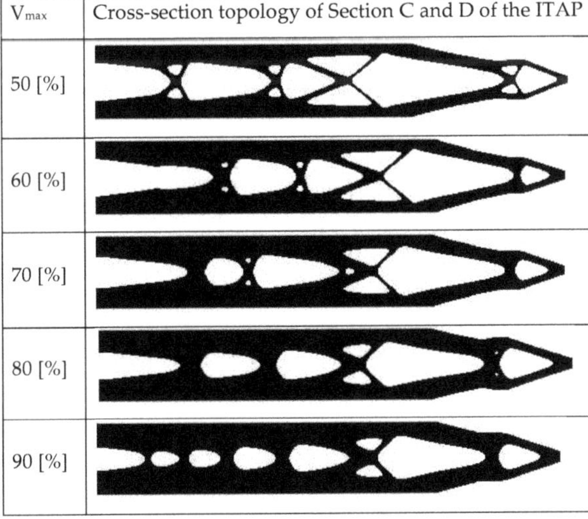

V_{max}	Cross-section topology of Section C and D of the ITAP
50 [%]	
60 [%]	
70 [%]	
80 [%]	
90 [%]	

Figure 9. Simulated results of the 2D MATLAB topology simulations identifying the theoretical areas of material reduction of 50 to 95%.

3.3. 3D Simulations Using MATLAB Functions

To identify the influence of multiple forces on ITAP topology further simulations were needed. The work of Liu et al. allows for simple 3D models to be designed and simulated where fixed support and applied loads can be controlled with multiple forces being utilised at a single location [23]. Using cubic models three separate simulation models have been studied looking at Sections C and D of the ITAP. These consist of a 85 mm by 14 mm by 14 mm model (Standard), as well as two models with cross-sections through the centre of the model at 85 mm by 14 mm by 1 mm (Vertical) and 85 mm by 1 mm by 14 mm (Horizontal). A 3D multsicale topology optimization code for lattice microscale response is used (top3d, Nelx, Nely, Nelz, Volfrac, Penal, Rmin) and the assumptions made for this study are as follows:

- The dimensions of the geometry are individual to each simulated model; this consists of Nelx, Nely and Nelz.
- Volfrac represents the volume fraction limit of the simulations; the range used for this study will be concordant with the 2D MATLAB study ranging between 50% and 95% in intervals of 5%.
- Penal represents penalization procedure and is usually referred to as the Solid Isotropic Material with Penalization (SIMP), which is used as a binary solver where penal > 1 In this study it is being set as a constant of 1 following the recommendations of the author.
- Rmin is the filter size which is the distance between the centroid of element i and element j of the model, this has been set to a value of 1, so there is a millimetre between the centre of each element.
- The Young's modulus of the material has been set to 1 while the Young's modulus of the voids has been set to the value of 0.
- Poisson's ratio is set to 0.3.
- The fixed supports are distributed evenly over the following coordinate positions fixed in all three axes:
    - Standard: X1 = 0, X2 = 0, Y1 = 0, Y2 = 14, Z1 = 0, Z2 = 14.
    - Vertical: X1 − 0, X2 − 0, Y1 − 0, Y2 − 14, Z1 − 0, Z2 − 1.
    - Horizontal: X1 = 0, X2 = 0, Y1 = 0, Y2 = 1, Z1 = 0, Z2 = 14.
- The applied forces are Axial 3168.63 N, Lateral −833.85 N and Frontal 958.93 N in the following axes Fx, Fy, Fz, respectively.
- The following coordinates and forces have been used in the load distribution:
    - Standard: Fx, Fy and Fz at X1 = 85, X2 = 85, Y1 = 0, Y2 = 14, Z1 = 0, Z2 = 14.
    - Vertical: Fx and Fy at X1 = 85, X2 = 85, Y1 = 0, Y2 = 14, Z1 = 0, Z2 = 1.
    - Horizontal: Fx and Fz at X1 = 85, X2 = 85, Y1 = 0, Y2 = 1, Z1 = 0, Z2 = 14.

The MATLAB 3D simulations provide the most detailed results. From the study, the areas where material can be removed to save weight due to the minimal σ present is consistently located at two positions. These being at the centroid of the fixed support at the boundary wall where the ITAP's Sections B and C meet. The next area where material can be removed is located within Section D, 57 mm < x < 85 mm see Figure 10A, where material is being removed on the opposite side of the applied Lateral and Frontal loads. These two positions of material reduction are concordant with both the ANSYS and 2D MATLAB simulations as well as all three models of the 3D MATLAB simulations. The size of the voids forming is dependent on the volume percentage of material being simulated. The results from the Standard models identify how the ITAP could be shaped to meet the individual's body characteristics using the loads calculated. The results identify the areas of material that are less important to the structural integrity of the external sections of the ITAP. The Vertical and Horizontal simulations look at the cross-section of the ITAP with the Axial and either Lateral or Frontal forces applied, Furthermore, the remaining internal structure mimics the internal structure found in bone, Figure 10, forming pathways in-between voids in the ITAP, similar to the findings of both Otomori and Wu with their respective studies and coding [22,24]. Both the Vertical and Horizontal results show this pattern. The voids are geometrically triangular in design as they are extremely capable of distributing σ within their shape [24]. There is minimal variation between the two sets of results, and this is expected as the forces being applied are similar in size. As the percentage of the material used increases, the definition of the voids and interconnecting pathways decreases. For these study 30 simulations were conducted utilising 10 different volume fraction limits for three different tested geometries.

The visual results of the simulations can be seen in Figure 10 for selected volume percentages of material.

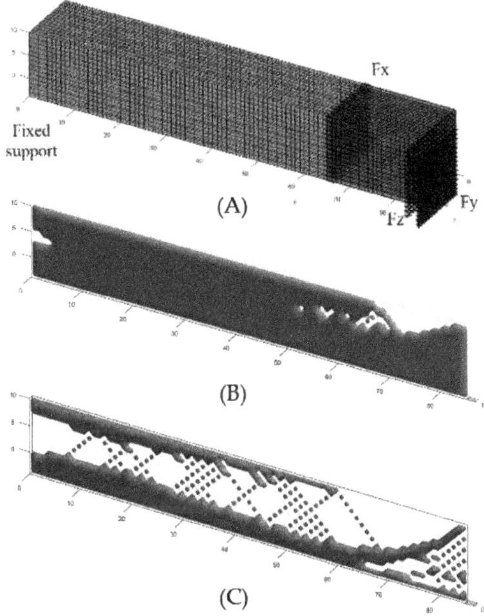

Figure 10. (**A**) FBD of the 3D MATLAB topology simulation utilising the external Sections, C and D, of the ITAP, dimensions of 85 mm by 14 mm by 14 mm with applied loads, (**B**) Vertical model using F_x and F_y 95% volume of material, (**C**) Vertical model using F_x and F_y 55% volume of material.

4. Topology of the ITAP Design under Extreme Loads

In this study the shear force being applied is set to the interconnecting surface between the prosthetic and the ITAP, represented by F_y as −4000 N. There is a secondary axial force of 416.9 N applied, denoted by F_z. This has been derived from half of the user's weight resting on the ITAP, from the 85 kg test individual. From analysing the results of the simulation, the location of the σ_{Max} can be seen to be at the joint between Section A and B, where the ITAP leaves contact with the bone anchor, as seen in Figure 8B. This is the same location as in the walking simulations. The only difference is the position on the circumference of the joint of Sections A and B, as the forces are acting in different directions. The σ_{Max} has been recorded above the σ_s of Ti and both the compressive and tensile σ_s of bone; therefore, a fracture has a high chance of occurring under this load. From these findings, it is recommended that further study is needed into the design of the ITAP to remove the risk of critical failure of the bone when exposed to extreme loads.

4.1. ANSYS Topology of the ITAP with Extreme Loads Applied

For this study the assumptions and controls are the same as in Section 3.1. The results identify that voids are forming in the same locations as in the walking simulations, these being at the boundary walls between sections and fixed supports. Material is also being removed along the sides of the ITAP in the same manner as the walking simulations. However, the main differences in the results is the voids forming within Section D, where more material remains in these locations to disperse the σ throughout the model from the applied loads. As such 4-separate standard ITAP models where simulated utilising ANSYS's high computational topology modelling process. As seen in Figures 11 and 12.

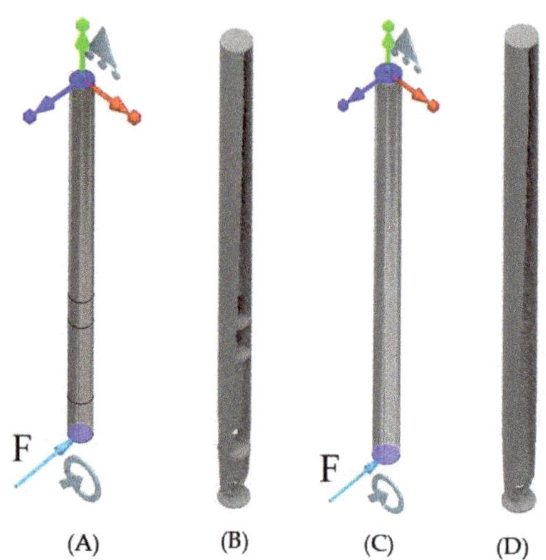

Figure 11. Topology simulations of the full standard ITAP with extreme load F, which is the amalgamation of F_y and F_z at 50% volume of material, (**A**) FBD of the ITAP in Sections, (**B**) The resulting topology of the individual Sections, (**C**) FBD of the ITAP constructed in one Section, (**D**) The resulting topology of the full ITAP.

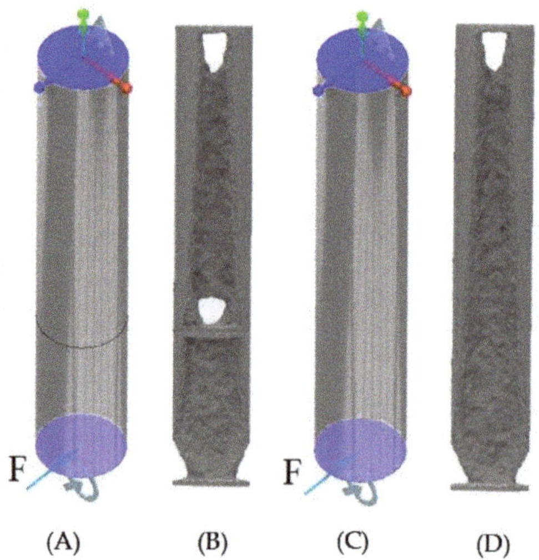

Figure 12. Topology simulations of Sections C and D of the standard ITAP with extreme load F, which is the amalgamation of F_y and F_z at 50% volume of material, (**A**) FBD of the ITAP in Sections, (**B**) The resulting topology of the individual Sections, (**C**) FBD of the ITAP constructed in one Section, (**D**) The resulting topology of Sections C and D.

4.2. 3D MATLAB Analysis of the ITAP with Extreme Loads Applied

For this study the assumptions and controls are the same as in Section 3.3 here the geometries used are the Standard and Vertical models. The simulations have been run between 50 and 95% of material in intervals of 5%, resulting in 30 separate geometries being developed for simulations. This study has identified the critical areas of material within the ITAP's Sections C and D at different volume percentages of material, by removing material that had the lowest σ concentration within the geometry. Common trends have been identified in these simulations. This has included: voids forming at the centre of the fixed support; divots manifesting within Section D; triangular interconnecting pathways forming within the voids in the centre of the models; and critical material kept along the boundary walls, most prominently the wall with the force directed towards it. All these observations are identical to the walking simulations 3D MATLAB study. As the percentage of the volume increases, the voids decrease in size as expected, but the pattern built into the structure is constant with the exception of the vertical model at 95%. The results from the latter simulation is sporadic because the material removed is not concordant with previous iterations.

5. Discussion

5.1. Similarities of the ANSYS, 2D and 3D MATLAB Topology Studies on the ITAP for Walking Simulations

Each simulation process has identified which regions of material are critical for the distribution of σ within the ITAP and what material can be removed to reduce the mass of the ITAP. As stated before, these results identify the material that can be removed as it is not critical to the structure of the ITAP. As such, the inverse of the results should be studied to correctly design the failure features of the ITAP, so that the device can be correctly used in day-to-day life, but will fail in a designed manner when exposed to extreme loads. Looking at Figure 13 which shows the visual results for each simulation method for sections C and D of the ITAP at 50% by volume of material, a clear pattern can be identified for further study. There are similarities in the reduction of material in each process, most notably the reduction of a large proportion of material at the fixed support of Section C, which is present in all models. Other significant similarities are the reduction of material identified in Section D, where material is removed in the opposite direction of the Frontal and Lateral forces for the divot. This is most prominently observed in both the ANSYS and 3D MATLAB simulations as the magnitude of the forces has been directly controlled and adjusted accordingly. The inverse of the results shows that the σ found within these locations is not substantial and therefore not critical for material reduction. The main areas where reduction could be focused on to achieve a controlled failure are the two quarters of the models that do not have any change in design; this can most prominently be seen in D in Figure 13 as well as A and B. Both the ANSYS and the Standard 3D MATLAB simulations are very similar in the locations of material reduction; the main differences arise due the starting geometry used.

The 2D MATLAB simulated model C is the most diverse, because the algorithm calculating the theoretical structure that could be capable of withstanding shear forces only at a point location rather than a dispersed load that is controlled. As such, the 2D MATLAB simulations do not work out the best structure to an individual load parameter, but rather work out the optimum geometry for shear loads being applied.

The results from this study have identified which internal structures should be utilised to distribute σ while saving weight. There are similarities in the simulated models such as the aforementioned void forming at the top of Section C in the ITAP, as well as the voids in the main body of the section being triangular based; this is similar to the results of E and F in Figure 13. However, from analysing and inverting the results provided from the use of Otomori's code, the areas of critical importance for distributing σ when designing the internal topology for shear σ are the boundary walls of the model. This is concordant with the range of percentages of volume simulated in this study. The resulting design structures (Figure 9) show that as the maximum volume of the model increases, the size of the voids forming in the ITAP decrease. This is most prominent in the voids forming at the boundary wall and the voids forming where sections C and D meet. This study has identified where the areas of the topology of the ITAP should be changed depending on the volume of the material used in additive manufactured production processes for complex internal structures. The topology simulation using MATLABs' applied load only consisted of one shear force, rather than the three forces that are applied to the ITAP in motion. Furthermore, the force cannot be set to the applied loads identified in Section 2.1 As such, the results from this study can only be used as a guide. Bye analysing the studies it can be seen that only critical material is left, thus it can be inferred that this is the area where design changes are required for a controlled failure. This is concordant with the work of Bird et al. [13], where the location and depth of the safety notches affected the ITAP performances.

The findings from the 3D Vertical and Horizontal MATLAB studies support this claim. However, the findings identify that one section of the material has more importance than the other. In Figure 13 both E and F's cross-section of the ITAP has a thicker boundary wall on one side, located in the direction of F_y or F_z, respectively. Both these locations are more regular and thicker than the opposite side of the ITAP. Inferring that the thicker side is more important for structural integrity and therefore should be the focus location when designing in the controlled location of failure. In total 60 different simulations were conducted to identify and consolidate the critical material needed for the structural stability of the ITAP when simulated with walking loads.

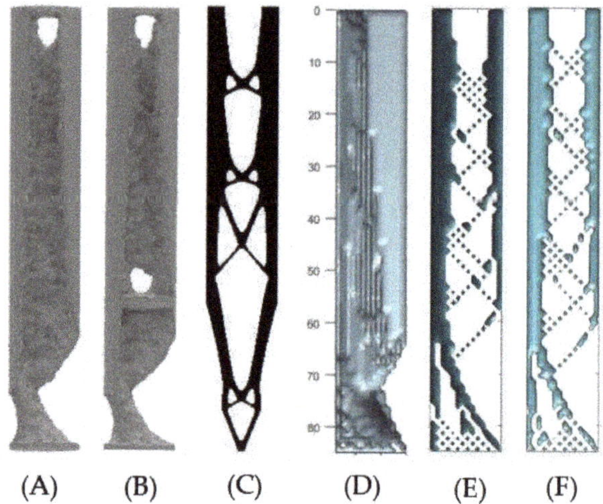

Figure 13. Visual comparison between topology simulation of Section C and D of the ITAP at 50% volume of material, (**A**) one solid model using ANSYS's control, (**B**) 2 sections joined together using ANSYS's simulation controls, (**C**) 2D MATLAB simulations with only the shear force being applied to identify the potential areas of material reduction, (**D**) 3D MATLAB standard model, (**E**) 3D MATLAB vertical model, (**F**) 3D MATLAB horizontal model.

From the results of these studies (Table 2), the following conclusions and recommendations can be made for further development in simulation and physical testing of the topology of the ITAP:

- Integrate a stress raiser on the ITAP at the end of Section B at the bone anchor, with appropriate fillets on either side to increase contact with bone and reduce residual σ under load at the connection between the ITAP and anchor.

- The external walls around the ITAP are key to the stability of the design; minimal removal in any direction will weaken the model. However, removing critical material from the combined direction of the Lateral and Frontal forces will introduce a significant weakness in design. Removal of material in the opposite direction will also have a significant reduction in strength.
- The internal topology of the ITAP should be designed following the results of the study's triangular-based patterns for stability, with AM production processes in mind.
- Any design models should be simulated using the forces utilised in these studies as well as simulation with forces that would fracture the femur.

Table 2. Simulated σ_{Max} and mass reduction of the ITAP when studied at 50% topology reduction.

σ_{Max} and Mass Reduction	Walking Loads				Extrema Load
ITAP Designs	**Standard**	**13 mm Safety Notch**	**9 mm Safety Notch**	**5 mm Safety Notch**	**Standard**
σ_{Max} [MPa] In the model	1175.6	939.5	939.5	3313.8	3220.9
σ_{Max} [MPa] Core of the ITAP	79.5	74.5	146.1	1003.6	211.7
Vertical height from applied loads, z domain [mm]	104.5	104.5	55.1	55.0	104.5
σ_{Max} [MPa] in Section A	570.3	735.6	735.6	735.6	1813.7
σ_{Max} [MPa] Core of Section A	67.1	72.1	72.1	72.1	215.0
Vertical height from applied loads, z domain [mm]	107.4	107.4	107.4	107.4	107.4
σ_{Max} [MPa] in Section C	422.6	474.5	861.6	3313.8	1264.0
σ_{Max} [MPa] Core of Section C	29.5	30.8	150.3	895.4	64.8
Vertical height from applied loads, z domain [mm]	64.7	49.9	53.3	59.0	69.2
Mass of full ITAP [g]	187.8	187.7	186.7	184.5	187.8
Mass of full ITAP after 50% topology reduction [g]	93.91	93.8	93.3	92.2	93.9
Mass of Section C & D of ITAP [g]	60.5	60.4	59.4	57.2	60.5
Mass of Section C & D of ITAP after 50% topology reduction [g]	30.2	30.2	29.7	28.6	30.2

5.2. Discussion and Similarities of the ANSYS and 3D MATLAB Topology Studies on the ITAP when Exposed to Extreme Loads

From analysing both the data from ANSYS and 3D MATLAB topology studies for an ITAP exposed to extreme loads, see Figure 14, that are expected to cause fracture to the femur, the following observations and conclusions can be drawn:

- The σ within the ITAP is significantly greater than the capacity of the anchor. Therefore, design iterations are needed to remove the failure of bone, and the location of the σ_{Max} is the same as the walking simulations in the anchor.
- The external walls, perpendicular to the forces are critical to the structural integrity of the ITAP. This is more prominent in the wall to which the shear force is directed.
- Voids in the centre are not critical to the structure but there is a triangular interconnecting pattern similar to the structure of bone which is required for structural stability within the models [21,23,24].

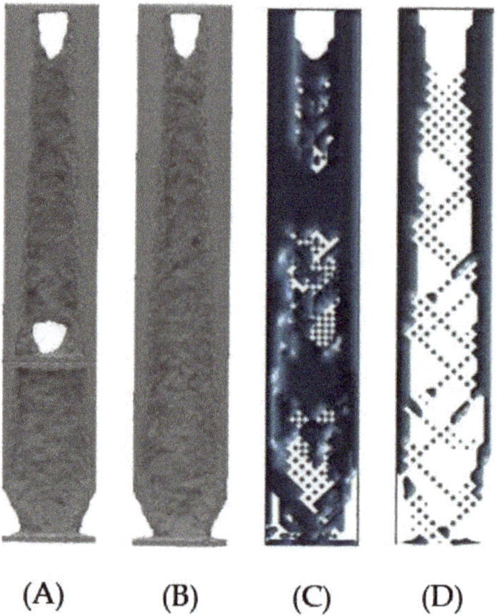

Figure 14. Visual comparison between topology simulation of Sections C and D of the ITAP at 50% volume of material, (**A**) one solid model using ANSYS's control, (**B**) 2 Sections joined together using ANSYS's simulation controls, (**C**) 3D MATLAB standard model, (**D**) 3D MATLAB vertical model.

The findings from these studies are identical with the walking simulations, indicating that critical material should be removed from the model on the boundary wall, to which the shear the force is directed. However, the difference in the positioning of the forces is subject to change for the extreme loads because impact could occur from any direction. Thus, a failure feature must be designed that is capable of failing from any direction of shear force. This was concordant with all 34 simulations. The similarities of the location and σ_{Max} recorded for both the walking and extrema loads can be seen in Table 2. This also identifies the mass reduction calculated from the ANSYS calculations.

6. Conclusions and Recommendations for Future Work

The ITAP is an alternative prosthetic product that has the potential to remove the inflammation and pain from traditional prosthetics while replicating the feeling of having a leg via a direct skeletal implant, thus improving the quality of life of the user. The inherent risk of using an ITAP is that because it is attached to the bone any σ on it receives is transferred to the bone. To prevent further limb damage a failsafe safety notch design is considered. The optimisation challenge is to ensure the ITAP is functional for normal use while allowing for a designed failure if excessive forces are experienced. The following conclusions are made.

- Due to advances in medical practises, CT scanning has been performed identifying the condition and strength of the bone anchor. Thus, allowing for clear understanding of how an ITAP will integrate within the user after an amputated femur. The acting loads applied to the ITAP when in use, have been ascertained, allowing for design and manufacture using AM methods to incorporating topology modifications bespoke to the users' lifestyle.

- The topology studies of the ITAP have been performed using ANSYS and 2D and 3D MATLAB models. The use of the 2D modelling has quick results that allows for rapid modelling, with alternative length of prosthetics. However, there is no control on the magnitude of the applied load. As such it is only recommend for quick analysis for an idea of where material is critical with in a cross section. The use of the 3D code is better suited to live modelling as there is more control on both the material properties and applied loads. The benefit of this is a visual guide to the critical material that is accurate to the model. However, the problem is the simulation does not yield numerical results and have limited geometry modelling. As such it is best suited for quick accurate visual modelling of cross-sections. In total 10 2D and 60 3D Matlab simulations where carried out within this research, all results yield from these simulations gave clear insight into the topology of the ITAP under load.

- ANSYS code has been developed for advances topology modelling. This has allowed for in-depth studies into the ITAPs geometry looking at the core material and σ that are being applied to the ITAP. This has allowed for in-depth review of the ITAPs critical material and local maximum stresses when modelling the applied loads. The downside of this simulation is that it is computationally expensive and time consuming, a large proportion of time was utilised simulation the 24 different ANSYS topology models.
- The topology studies for both the walking and extreme loads have identified critical areas of material within the ITAP that are needed for σ to dissipate through the product. Through the use of ANSYS, 2D and 3D MATLAB models, it can be concluded that when the ITAP is exposed to walking loads, the critical material is solely located in line with the summation of the walking loads, which is set in a fixed direction. It is also critical that the section of the ITAP that meets the bone anchor does not exceed the bone σ_s of 110 MPa as this will cause further injury to the user.
- The critical support material identified from the topology studies is directly in line with the extreme loads. However, in use as an extreme load can be applied from any direction, and all of the ITAP geometry can be assumed as critical so a controlled failure safety notch feature is adopted. As such it can be stated that a developed ITAP must not exceed the σ_s of bone but also be capable of withstanding loads for daily use like walking. For the failsafe to work the σ within the core section C must exceed the material UTS when exposed to extreme loads in any direction.

It is recommended that further research is undertaken on the design of the ITAP's failure feature utilising the topology study in this paper so that a safety design can be successfully developed for each individual case. Physical testing is also highly recommended as to ascertain the strength compression on different volume fraction of the ITAP, when tested with walking loads and destructive forces. Further simulation on different loads including running and jumping to fully facilitate the user's mobility would gain an insight in the loading capabilities of the ITAP.

Author Contributions: E.L.; formal analysis, investigation, software, visualization, methodology. C.A.G.; supervision, conceptualization, methodology, writing—original draft preparation. A.R.; resources, data curation, writing—review and editing. J.B. validation, formal analysis, visualization. All authors have read and agreed to the published version of the manuscript.

Funding: This research received no external funding.

Acknowledgments: The authors would like to acknowledge the support of the Future Manufacturing Research Institute, College of Engineering, Swansea University and Advanced Sustainable Manufacturing Technologies (ASTUTE 2022) project, which is partly funded from the EU's European Regional Development Fund through the Welsh European Funding Office, in enabling the research upon which this paper is based. Further information on ASTUTE can be found at www.astutewales.com (accessed on 11 March 2021).

References

1. Charles, Q. Brute Force: Humans Can Sure Take a Punch. *LIVESCIENCE*. 3 February 2019. Available online: https://www.livescience.com/6040-brute-force-humans-punch.html (accessed on 3 August 2019).
2. Teitel, A.S. How Much Force Does It Take to Break A Bone? *Seeker*. 20 October 2016. Available online: https://www.seeker.com/how-much-force-does-it-take-to-break-a-bone-2056139307.html (accessed on 3 August 2019).

3.   Mckay, B. Development of Lower Extremity Injury Criteria and Biomechanical Surrogate to Evaluate Military Vehicle Occupant Injury during an Explosive Blast Event. Bachelor's Dissertation, Wayne State University, Detroit, MI, USA, 2010; pp. 1–146.
4.   Langford, E.; Griffiths, C.A. The mechanical strength of additive manufactured intraosseous transcutaneous amputation prosthesis, known as the ITAP. *AIMS Bioeng.* **2018**, *5*, 133–150. [CrossRef]
5.   Zhang, M.; Turner-Smith, A.R.; Roberts, V.C.; Tanner, A. Frictional action at lower limb/prosthetic socket interface. *Med. Eng. Phys.* **1996**, *18*, 207–214. [CrossRef]
6.   Pendegrass, C.J.; Goodship, A.E.; Price, J.S.; Blunn, G.W. *Nature's Answer to Breaching the Skin Barrier: An Innovative Development for Amputees*; The Centre for Biomedical Engineering, Institute of Orthopaedics and Musculoskeletal Science: Stanmore, UK, 2006; pp. 56–67. [CrossRef]
7.   Niinomi, M. Mechanical properties of biomedical titanium alloys. *Mater. Sci. Eng. A* **1998**, *243*, 231–236. [CrossRef]
8.   Handbook of Materials for Medical Devices. Overview of Biomaterials and Their Use in Medical Devices. Available online: https://www.asminternational.org/documents/10192/1849770/06974G_Chapter_1.pdf (accessed on 4 July 2019).
9.   Asri, R.I.M.; Harun, W.S.W.; Samykano, M.; Lah, N.A.C.; Ghani, S.A.C.; Tarlochan, F.; Raza, M.R. Corrosion and surface mod-ification on biocompatible metals: A review. *Mater. Sci. Eng.* **2017**, 1261–1274. [CrossRef] [PubMed]
10. Deng, L.; Deng, Y.; Xie, K. AgNPs-decorated 3D printed PEEK implant for infection control and bone repair. *Colloids Surf. B Biointerfaces* **2017**, 483–492. [CrossRef] [PubMed]
11. Kang, N.V.; Morritt, D.; Pendegrass, C.; Blunn, G. Use of ITAP implants for prosthetic reconstruction of extra-oral craniofacial defects. *J. Plast. Reconstr. Aesthetic Surg.* **2012**, *66*, 497–505. [CrossRef] [PubMed]
12. Newcombe, L.; Dewar, M.; Blunn, G.; Fromme, P. Effect of amputation level on the stress transferred to the femur by an artificial limb directly attached to the bone. *Med. Eng. Phys.* **2013**, *35*, 1744–1753. [CrossRef] [PubMed]
13. Bird, J.; Langford, E.; Griffiths, C. A study into the fracture control of 3D printed intraosseous transcutaneous amputation prostheses, known as ITAPs. *AIMS Environ. Sci.* **2020**, *7*, 29–42. [CrossRef]
14. Nebergall, A.; Bragdon, C.; Antonellis, A.; Kärrholm, J.; Brånemark, R.; Malchau, H. Stable fixation of an osseointegated implant system for above-the-knee amputees. *Acta Orthop.* **2012**, *83*, 121–128. [CrossRef] [PubMed]
15. Fitzpatrick, N.; Smith, T.J.; Pendegrass, C.J.; Yeadon, R.; Ring, M.; Goodship, A.E.; Blunn, G.W. Intraosseous Transcutaneous Amputation Prosthesis (ITAP) for Limb Salvage in 4 Dogs. *Veter. Surg.* **2011**, *40*, 909–925. [CrossRef] [PubMed]
16. Sullivan, J.; Uden, M.; Robinson, K.P.; Sooriakumaran, S.; Robinson, P.K.P. Rehabilitation of the transfemoral amputee with an osseointegrated prosthesis. *Prosthet. Orthot. Int.* **2003**, *27*, 114–120. [CrossRef] [PubMed]
17. Mahmoudi, M. Femur Bone. Grabcad Community. Available online: https://grabcad.com/library/femur-bone-2 (accessed on 5 July 2019).
18. Rybicki, E.F.; Simonen, F.A.; Weis, E.B., Jr. On the mathematical analysis of stress in the human femur. *J. Biomech.* **1972**, *5*, 203–215. [CrossRef]
19. Marco, M.; Giner, E.; Larraínzar-Garijo, R.; Caeiro, J.R.; Miguélez, M.H. Numerical modelling of femur fracture and experimental validation using bone simulant. *Ann. Biomed. Eng.* **2017**, *45*, 2395–2408. [CrossRef] [PubMed]
20. Reilly, D.T.; Burstein, A.H. The elastic and ultimate properties of compact bone tissue. *J. Biomech.* **1975**, *8*, 393–405. [CrossRef]
21. Duda, G.N.; Schneider, E.; Chao, E.Y. Internal forces and moments in the femur during walking. *J. Biomech.* **1997**, *30*, 933–941. [CrossRef]
22. Otomori, M.; Yamada, T.; Izui, K.; Nishiwaki, S. Matlab code for a level set-based topology optimization method using a reaction diffusion equation. *Struct. Multidiscip. Optim.* **2015**, *51*, 1159–1172. [CrossRef]
23. Liu, K.; Tovar, A. An efficient 3D topology optimization code written in Matlab. *Struct. Multidiscip. Optim.* **2014**, *50*, 1175–1196. [CrossRef]
24. Wu, J.; Aage, N.; Westermann, R.; Sigmund, O. Infill Optimization for Additive Manufacturing—Approaching Bone-Like Porous Structures. *IEEE Trans. Vis. Comput. Graph.* **2018**, *24*, 1127–1140. [CrossRef] [PubMed]

MDPI

St. Alban-Anlage 66

4052 Basel

Switzerland

Tel. +41 61 683 77 34

Fax +41 61 302 89 18

www.mdpi.com

Micromachines Editorial Office

E-mail: micromachines@mdpi.com

www.mdpi.com/journal/micromachines